CAUSE, EFFECT, AND EVERYTHING IN BETWEEN

An Introduction to Causal Inference

Aboozar Hadavand

OXFORD
UNIVERSITY PRESS

OXFORD
UNIVERSITY PRESS

Oxford University Press is a department of the University of Oxford.
It furthers the University's objective of excellence in research, scholarship,
and education by publishing worldwide. Oxford is a registered trade mark of
Oxford University Press in the UK and in certain other countries.

Published in the United States of America by Oxford University Press,
198 Madison Avenue, New York, NY 10016, United States of America.

© Oxford University Press 2025

All rights reserved. No part of this publication may be reproduced, stored in a retrieval
system, transmitted, used for text and data mining, or used for training artificial
intelligence, in any form or by any means, without the prior permission in writing of
Oxford University Press, or as expressly permitted by law, by license or under terms
agreed with the appropriate reprographics rights organization. Inquiries concerning
reproduction outside the scope of the above should be sent to the Rights Department,
Oxford University Press, at the address above.

You must not circulate this work in any other form
and you must impose this same condition on any acquirer.

CIP data is on file at the Library of Congress

ISBN 9780197801789

ISBN 9780197801772 (hbk.)

DOI: 10.1093/9780197801819.001.0001

Paperback printed by Integrated Books International, United States of America

Hardback printed by Lightning Source, Inc., United States of America

The manufacturer's authorized representative in the EU for product safety is
Oxford University Press España S.A., Parque Empresarial San Fernando de Henares,
Avenida de Castilla, 2 – 28830 Madrid (www.oup.es/en).

The manufacturer's authorised representative in the EU for product safety is Oxford University Press España S.A. of El Parque Empresarial
San Fernando de Henares, Avenida de Castilla, 2 – 28830 Madrid (www.oup.es/en or product.safety@oup.com). OUP España S.A. also acts as
importer into Spain of products made by the manufacturer.

CAUSE, EFFECT, AND EVERYTHING IN BETWEEN

To Sarah

CONTENTS

Preface | viii

1. What Is Causality? | 1
2. The Causal Framework | 17
3. Causal Graphs and Causal Paths | 28
4. Causal Inference Using Interventional Data | 40
5. Causal Inference Using Observational Data | 58
6. Quasi-Experimental Methods | 75
7. A Framework for Evaluating Causal Studies | 97
8. Causal Case Studies | 109

ANSWERS TO END-OF-CHAPTER QUESTIONS | 124
INDEX | 136

PREFACE

When I was an undergraduate engineering student, I only took one introductory course in statistics and probability—and I nearly failed it. While I hold no ill will toward the instructor, the course didn't spark any interest in me. Late in my twenties, as I was getting my PhD in economics, I encountered econometrics (a fancy term economists use for causal inference). While I started to appreciate the value of causal inference, I still felt intimidated by it. The courses were heavy on esoteric terms and symbols and light on reasoning and intuition. As my students like to say, "It felt too math-y."

Fortunately, my PhD years coincided with a shift in academics' focus toward the intellectual foundations that take causal inference beyond the realm of statistical analysis. In social sciences, this shift was driven by the so-called credibility crisis, introduced by economist Edward Leamer in 1983. The credibility crisis in econometrics and other neighboring disciplines refers to a period during the 1980s when the reliability and validity of empirical economic research were widely questioned. Scholars expressed concerns that many econometric studies produced inconsistent or nonreplicable results, casting doubt on the dependability of empirical findings within the discipline. The issue was that researchers focused too much on choosing the right statistical model, assumptions, and hypothesis testing rather than being transparent about the causal dynamics governing their research topics. The credibility crisis put causal inference at the forefront of econometrics, and its impact is evident in today's curricula. In recent years, topics like causal graphs, confounding bias, and causal discovery have become integral to undergraduate and graduate courses.

Many of these courses assume a foundational knowledge of statistics, linear algebra, and probability theory. I don't argue that these topics aren't important for applied researchers, but I believe critical causal thinking doesn't always require deep statistical expertise. In writing this book, I stand by three convictions. First, everyone should have a basic understanding of causality—whether a first-year undergraduate or a graduate researcher, a journalist or a policymaker, or someone exploring a disease alongside their doctor. Second, the most crucial concept in causal inference is understanding confounding bias. And lastly, it's possible to grasp the idea of confounding bias without heavily relying on mathematical notation. You don't need to be a statistician or econometrician to think critically about causal arguments.

This book is as nontechnical as the subject allows. My main goal is to give you a tour of the most important concepts in causal inference while keeping it as accessible and readable as possible. Writing this book was challenging and required careful attention, but I found the task engaging and fun. The book grew out of courses I've taught at Minerva University and beyond, and without my curious students there, I wouldn't have been as eager to learn more myself. I specifically thank my student Joram Erbarth, whose feedback and help were important in writing this book.

A final note: If you are looking for a more advanced resource in causal inference, sadly, you are reading the preface of the wrong book. This book serves as an introductory text on causal inference. Once you've mastered the basics, many excellent resources can take you further if you want to learn more about causal inference.

Aboozar Hadavand
Atlanta, Georgia
October 2024

1

What Is Causality?

My great-grandfather, whom I never met, is said to have lived more than one hundred years. He lived so long that he outlived four spouses. Throughout his life, he was healthy. He never once caught a cold; he worked until his last breath. The key to his longevity? Yogurt! Or so the family lore goes. My great-grandfather lived off a steady diet of yogurt at every meal—sometimes accompanied by a slice of bread or a piece of meat—and when he lived to be a centennial, everyone around him concluded it was the yogurt that made him live so long.

As a kid, I was always skeptical of the yogurt claim, but whenever my mom told me to eat a vegetable I didn't like, I wielded the claim to my advantage. "*Didn't Dad's grandpa live to be a hundred without eating these vegetables? I'll just have some yogurt instead.*" My mother would fight back with her own causal artillery. "*Raw garlic will boost your immune system.*" "*The carrot juice will make your eyesight better.*" "*This ginger tea will cure your cold.*"

Little claims, like, "*Yogurt will make you live to be one hundred,*" are everywhere. They are in advertisements and in the news. They are on the pages of parenting books, are embedded in the scientific community, lie at the root of how we govern cities and countries, and influence our lifestyle choices. Anytime we pose or answer a question in the form of *Is X responsible for a change in Y?*, we employ the notion of causality.

$$X \to Y$$

$$\text{Yogurt} \to \text{Long life}$$

Formally, we can say that X causes Y if Y responds to changes in X. Computer scientist Judea Pearl, a leading expert in causal inference, defines cause and effect as follows: "X is a cause of Y if Y listens to X and decides its value in response to what it hears."[1]

As humans, we owe a lot to our innate curiosity about causality. Our interest in causality helped our ancestors navigate the flora and fauna of the woods and eventually led them to things like modern forms of agriculture, medicine, and industry. It has kept us from danger: touching fire burns our hands, lightning may shock us, a poisonous mushroom may kill us. A lot of our progress as humans is a result of a simple question being asked over and over again: *What if I do this, or what if I do that instead?*

Causal reasoning is a large part of what differentiates us from other animals. It is not that all other animals are incapable of drawing causal conclusions—both my dogs have a keen sense of what follows when I approach their treats, a tennis ball, or a leash—but humans are far more interested and adept at asking, *Should I . . . ?*, *Why did [this] happen?*, *What if I . . . ?*, *Is [this] because of [that]?* Causation interests us because it speaks to our agency. We know that if we can understand how the world works, we can shape the future to our benefit. Causation can also be found in fiction, especially science fiction. We have stories about the state of the world had the Germans won World War II; had the South won the American Civil War. A lot of these stories are *what-if* stories.

As humans, we constantly have to make hard decisions. Is it a good investment to open a restaurant? Should we go to college? Should we get vaccinated, and how often? Which political party is

---

1. Causal Inference in Statistics: A Primer, First Edition. Judea Pearl, Madelyn Glymour, and Nicholas P. Jewell. © 2016 John Wiley & Sons, Ltd. Published 2016 by John Wiley & Sons, Ltd.

better equipped to deal with the economy? Will too much government spending lead to inflation? Should we close our borders? As a society, we need to be better at having these discussions.

Here's the rub. Part of the struggle with causal reasoning is that answering a causal question is much harder than asking one. For every causal question we answer correctly, there are countless falsehoods and untruths we pass around as causal explanations. As good as we are at asking causal questions, and notwithstanding the tremendous progress that has resulted from asking them, for most of us, our ability to answer causal questions is, well, very rudimentary and full of flaws and fallacies.

Our desire for reliable causal explanations often conflicts with other motives. We tend to embrace or reject a causal explanation not because we think it's true but because it is convenient or inconvenient, sounds exciting or dull, comes from a source we consider reliable or unreliable, or works to our advantage or disadvantage. None of these is the same as embracing a causal explanation because we are confident it speaks the truth.

One of the logical fallacies we deal with is called the post hoc fallacy. The post hoc fallacy stems from the Latin phrase *post hoc ergo propter hoc*, which translates to "After this, and therefore, because of this." In our everyday use of causal reasoning, we often assume that if one thing follows another, the former causes the latter. This can work if the series of events happens to be something like, *I touched the hot stove, and now I've burned my finger*, but it might not work so well when the series of events are such as, *I cheered for Team A, and Team B lost*, or *She became vegan and lost a lot of weight*, or *Every time I invest in a stock, its price goes down*.

I invest in a stock ⟶ The stock price goes down

Another logical fallacy is the questionable cause fallacy or the causal fallacy. This fallacy refers to situations where a cause is incorrectly identified simply because it is usually associated with the effect. In the eighteenth and nineteenth centuries, for example, physicians trying to uncover the causes of certain personality traits concluded that people have different personality traits because the sizes of their brain regions are different. A protruding lump in the back, front, or side of the skull could indicate a caring disposition, a practical turn of mind, or even an inclination toward criminality. Phrenology, now understood to be pseudoscience, became so popular that Queen Victoria ordered a famed phrenologist to measure and read her children's heads.

If you think causal confusion is a thing of the past, think again. Remember the early weeks of the COVID-19 outbreak? Most of us thought wiping down every single item we bought at the grocery store with whatever cleaning supplies we had would keep us safe from the virus. We washed our hands like surgeons prepping for surgery, all while singing Happy Birthday to ourselves twice in a row. Some of the new behaviors we adopted were effective, and even some of the behaviors we adopted that were not effective were reasonable, especially in a new environment where there were many unknowns and plenty of caution was warranted. However, some of our confusion, behavior, and debates could have been curtailed if more of us had the language and skills to communicate through cause and effect.

This book is all about causal inference—the systematic and empirical study of causal relationships. These methods, while good enough to reveal some truths, are by no means fool-proof. Lucky for us, over the last century, there have been important scientific advances and new empirical approaches to causal reasoning. These approaches are based on advanced logic, probability, and statistics, but many of their lessons are within the grasp of a casual reader.

We will see a more systematic way of answering causal questions through well-done, large-scale, randomized experiments. We will also see that when randomized experiments are not feasible, we can still answer causal questions through studies known as observational studies.

My aim is to help spread the language of causal inference to a general audience and to help you bring your causal reasoning skills to the next level. If we are better at causal inference, we can make better decisions, understand (and to some extent assess) scientific research, and be better citizens.

What Is Causal Inference?

Einstein called causal inference one of Western science's two most significant achievements. David Hume called it "the cement of the universe." Neither of these men was referring to causal claims like my family's yogurt hypothesis or the causal claims we find in infomercials and talk shows. Instead, Einstein and Hume were talking about a more systematic approach to causal reasoning.

We said earlier that X is a cause of Y if Y responds or "listens" to X.

$$X \longrightarrow Y$$

Before moving further, let's add a few more qualifiers to our definitions of cause and effect. First, if X is a cause of Y, it does not have to be its only cause. For example, blood pressure responds to sodium levels, so sodium is a cause of blood pressure. Sodium, however, is not the only determinant of blood pressure. Blood pressure listens to many other variables, like family history, tobacco use, and obesity.

Second, if X is a cause of Y, you might be tempted to think that Y is bound to happen if X happens. Here is where our definitions of cause and effect differ slightly from how causality is sometimes defined in philosophy or our everyday discussions. If you assume that a cause leads to an effect 100 percent of the time, you are using a deterministic definition of a cause—*if A occurs, then B must occur*. In this book, we're going to define causes in a probabilistic way. If X *tends* to cause Y, X is a cause of Y. If dropping a glass on the floor *can* lead to it breaking, dropping the glass on the floor is a cause of broken glass, even if the glass doesn't always break. Likewise, high sodium levels are a cause of high blood pressure, even if some patients with high sodium levels do not have high blood pressure.

In causal inference, the word *causal* refers to an analysis based on causes and effects. The word *inference* is the part that makes causal inference a scientific approach to studying causes and effects. Inference is an empirical conclusion based on evidence and reasoning. So causal inference is an empirical approach to causal reasoning. We are using the word inference to distinguish between our everyday use of causal reasoning and the more systematic approach to causal reasoning that has been evolving over the past century.

Because we are using a probabilistic and empirical approach to cause and effect, causal inference involves not only *identifying* causal relationships but *quantifying* them as well. In the second half of this book, we'll begin to see how experimental and observational data can help us find and estimate causal effects.

Causal Graphs for Mapping Causal Relationships

You may have noticed already that we have been using arrows. We use these arrows to show causal relationships between variables. We use causal graphs that consist of nodes and arrows to show more complicated causal relationships when more than two variables are involved. We learn about these graphs, also called directed acyclic graphs (DAGs), in chapter 3, but here is a quick prelude.

A causal graph or DAG shows variables using circles or nodes and causal relationships using arrows. The arrow originates at the causal node and points toward the affected node. The DAG on the following page shows two variables, D and Y, where D is a cause of Y.[2] The arrows in a DAG signify that causal relationships have a clear direction that is usually unidirectional. If D is a cause of Y, Y is not typically a cause of D.

---

2. This is what we mean by a variable: A variable can contain numeric or categorical values. For instance, a vector containing the heights of students in a classroom is a variable. Similarly, a vector containing the gender of all employees in a company is also a variable. A variable is a descriptor that can take multiple values. In this book, we use the term *variable* a lot.

Going back to the sodium and blood pressure example, the causal graph includes two main nodes: sodium intake and blood pressure. We know that the sodium level causes high blood pressure. Because D is a cause of Y, Y relies on D for its value. Then the arrow in the causal graph goes from sodium to blood pressure pointing to the fact that sodium intake affects blood pressure and not the other way.

We also said that sodium intake may not be the only cause of high blood pressure. There may be other nodes on the graph that lead to the node representing blood pressure. Tobacco and obesity can also contribute to high blood pressure in addition to sodium levels. For pretty much every effect, there could be multiple contributing causes.

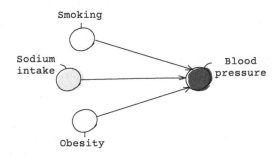

Does having high blood pressure (Y) have to be the only effect or consequence of sodium intake (D)? Also, no. D can cause other things that we may or may not be interested in; high sodium levels can cause other things in our bodies, but in our example, we were interested in blood pressure as the outcome.

So to wrap up, sometimes a cause has multiple effects, or an effect has multiple causes. A cause does not have to be necessary or sufficient. The sodium level is not sufficient nor a necessary contributor to high blood pressure.[3] Causation is also usually one way. If D causes Y, Y usually cannot cause D.[4] If happiness causes a dog to wag its tail, we do not expect the tail wagging to make the dog happy in return.

Hopefully, by now, you're beginning to get a sense of what causal inference is all about. In the next chapter, we talk about the *language* of causality. Chapter 3 helps you become a DAG master. In chapters 4 and 5, we start looking at causal inference using different types of data. Chapter 6 goes over a group of methods in causal inference called quasiexperimental methods. I am aware that devoting only one chapter to these methods may seem a bit rushed. My aim here is to at least introduce these methods at a surface level. The final two chapters of the book will prepare you to take your causal inference skills with you out into the world, so you can make better sense of political arguments, the news, medical studies, environmental studies, and the arguments thrown at you by friends and family.

Throughout this book, we will go through plenty of examples and case studies to help let the learning take hold. We start right here with four very short case studies to prepare you for understanding what causal inference is all about.

---

3. *Necessary* means unless D happens, Y cannot happen; *sufficient* means if D happens, Y always happens. This is not the case with sodium as the cause and blood pressure as the effect.

4. However, you might think of some counterexamples. For instance, you could argue that having more money leads to higher levels of education, and being more educated leads to more money. But you are missing the point that having more money now (let's say D at time t) may lead to more education now (Y at time t), and having more education now (Y at time t) may lead to more money in future (D at time t+1). So D at time t causes Y at time t, and Y at time t causes D at time t+1, but D at time t and D at time t+1 are not the same.

Case 1: Books and Income (Correlation Is Not Causation)

Consider the following data drawn from a sample of ten randomly selected individuals. The data tells you how many books people own and their salary. Pay attention to the data, and see if you can spot a trend or a causal relationship.

TABLE 1.1 Sample of ten students

Person ID	# of Books	Earnings
1	10	30,000
2	25	51,000
3	43	74,000
4	12	40,000
5	59	93,000
6	125	145,000
7	20	60,000
8	5	32,000
9	44	67,000
10	38	45,000

You'll notice that those who own more books tend to make more money. If we plot the data on a scatter plot, it looks something like the graph below. We can see that those with the fewest books have the lowest incomes, and those with more books have higher incomes. The richest person in the bunch has a far bigger book collection than the rest.

Based on this data, can we infer that owning books causes higher earnings? We can't. Lesson number one in causal inference is that *correlation is not causation*. You might have heard this phrase. It means that even if we find an association between two variables, we cannot conclude that one causes the other. For example, we might

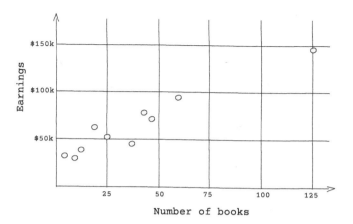

FIGURE 1.1 The scatter plot of earnings versus the number of books for the sample data

have some theory that reading more books makes you smarter, and smarter people tend to have higher earnings. That could be the case, but it could also be the case that older people tend to have more books (by accumulation) but also tend to earn more.

So association does not necessarily mean that one factor causes the other. Distinguishing between association and causation is one of the themes of this book. We will learn that if two variables move together (in the same or opposite directions), one is not necessarily caused by the other.

We'll also learn that association can be measured by an index called the correlation coefficient. A correlation coefficient is a number between −1 and +1 that captures how two variables move with each other. If the coefficient is close to zero, the association between the two variables is small. If the correlation coefficient is close to +1, the association between the two variables is strong and positive (the two variables move in the same direction). If the correlation coefficient is close to −1, the association between the two variables is strong and negative (the two variables move in opposite directions).

Since an associative relationship is agnostic to causality and its direction, in a DAG we show correlations by adding a nondirectional link between the nodes, like the DAG earlier. Note that there is no arrow in the figure since correlations do not have a direction.

Case 2: A New Drug for Asthma (Confounding Bias)

Next, consider a medical study in which researchers are trying to test a new asthma treatment. In other words, they are interested in studying the treatment's causal effect on asthma recovery.

The researchers randomly selected 1,000 people who have asthma who volunteered to be part of the study. The researchers want to divide these participants into two groups: those who receive the treatment and those who will not. They decide to leave the choice up to each individual. Of the group, 550 people choose to take the drug, and 450 choose not to. At the end of the study, 5 percent of the group receiving the drug saw improvement in their asthma, and 10 percent of those who did not receive the drug saw improvements. What, if anything, can we say about the causal relationship between the treatment and recovery from asthma?

The data suggest that those who received the drug had a lower recovery rate, which might lead us to believe that the drug may be harmful rather than effective. But be careful; the data here can be misleading.

The researchers of this hypothetical study made one major misstep while designing their experiment. They let the patients decide which group to be in. The approach of allowing participants to choose which group to be in immediately tarnishes the results.

Similar to case 1, what if other variables are involved? For example, what if it was the case that those who were more in need of the drug because they had a more acute form of asthma were more likely to opt in to the group that received the drug? Furthermore, what if those with more acute asthma were much less likely to see improvement in their asthma whether they received the drug or not? If this were the case, the drug could be effective, but we would be blind to the effect because of the composition of the treatment and control groups.

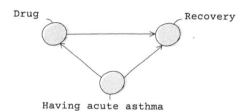

The results do not tell us whether the drug causes a lower or higher recovery rate. We need better techniques to answer questions like this. We will later see that this is a case of confounding bias. Confounders are a thorn in the side of causal inference. If the researchers want to make inferences from experiments like this, as we learn later, they need to make sure that participants are randomly selected for the study but also randomly assigned to the treatment and control groups.

Case 3: Swimmer's Body Illusion (Selection Bias)

Have you ever wondered why swimmers' bodies are so fit? We may think that training and exercise are what shape a swimmer's body, but in fact, it's the opposite. Individuals shaped like swimmers are more likely to be swimmers in the first place. They become competitive swimmers because they have a body suited to swimming. This problem is something we aptly call the swimmer's body illusion.

Marketing firms use the swimmer's body illusion to trick us into buying many things. You see superattractive models wearing brands of clothing or driving fancy cars, and it seduces us into thinking that if we wear the clothes or drive the car, we will be as attractive as the models. You can see examples of this in academia too. The top universities in the world, like Harvard, Yale, and Oxford, usually create the illusion that they are places where genius is formed, but a lot of evidence shows that being smart in the first place is what gets you admitted to the school. In other words, cause and effect may be working in a direction opposite to what we're led to believe.

Whenever you suspect a swimmer's body illusion, you need to see if there's a way to have a more representative sample[5] in your treatment and control groups. Rather than only looking at competitive swimmers, attractive people, or brilliant students, you need to see if a sample of people representing the population is experiencing the causal effects in question.

Later we'll see that the swimmer's body illusion is a case of sample selection bias. Sample selection bias happens when the sample we use to draw causal inferences is not representative of the population we want to study. If we are interested in the effect of swimming on body fitness, instead of using highly fit people as our sample, we should use a sample that better mimics the general population in which we are interested. We learn about sample selection bias in future chapters.

Case 4: California's Department of Developmental Services (Simpson's Paradox)

Here's the last case study for this chapter. Most states in the United States provide programs to support those with developmental disabilities, such as intellectual disability, cerebral palsy, and autism,

---

5. In statistics, a representative sample is a subset of a population that accurately reflects the characteristics and diversity of the entire population. A representative sample ensures that the conclusions drawn from the sample are generalizable to the whole population.

usually in the forms of financial and non-financial aid. California is one of the states that provide these benefits.

California allocates funds to support more than 250,000 developmentally disabled individuals through the Department of Developmental Services (DDS). Recently, however, DDS was accused of discrimination. Data suggested that, on average, a Hispanic resident with disabilities received $11,066, whereas their non-Hispanic White counterparts received an average of $24,698. The difference was considerable.

TABLE 1.2 Average expenditures and percentage of consumers per ethnicity

Ethnicity	Average Expenditures	% of Consumers
Hispanic	$11,066	38%
White non-Hispanic	$24,698	40%

DDS hired Stanley A. Taylor, a statistician from California State University in Sacramento, to investigate the claim. Since critics accused DDS of using ethnicity to determine how much recipients would receive, the problem was a causal inference problem. Were they, in fact, doing this?

Taylor looked at the DDS data and started breaking it down by other variables, such as gender and age. He thought these other factors might also be in play. When he divided the data up by these other variables, the discrimination disappeared. In fact, the odds of receiving more funds slightly favored Hispanics. The data from his analysis are shown in the Table 3.

If you compare the amount of funding received by Hispanics and non-Hispanic Whites, you notice that for most age groups, the funding is higher for Hispanics. In all but one of the age groups, the trend is reversed. So which analysis can we trust? The one based on the aggregated data or the one based on the disaggregated data?

TABLE 1.3 Average expenditures and percentage of consumers among age groups

Age Cohort	Average Expenditures (Hispanic)	Average Expenditures (White Non-Hispanic)
0—5	$1,393	$1,367
6–12	$2,312	$2,052
13–17	$3,955	$3,904
18–21	$9,960	$10,133
22–50	$40,924	$40,188
51+	$55,585	$52,670

This case exemplifies Simpson's paradox, a situation in statistics and causal inference where aggregated data show one trend and disaggregated data show another. In this particular example, we can't immediately tell if there was, in fact, discrimination. The data leave us confused. Later, we dive deeper into this example and discover that Simpson's paradox is actually another case of confounding bias. We'll determine whether, in this case, age is a confounding variable or not. Hopefully, by the end of this book, Simpson's paradox will no longer be a paradox for you.

You're now ready to embark on your causal inference journey. The examples you've just seen are among some of the simpler questions that causal inference can help answer. By the end of this book, you'll have mastery over far more complex ones. You'll be drawing causal graphs with ease, explaining to your friends and colleagues how randomized experiments work, and you'll have developed an eye for spotting confounders and other types of biases. Here we go!

Chapter 1 Problems

1. What differentiates causal inference from our everyday use of causal language?

2. In causal inference, under what circumstances is X considered a cause of Y?
3. How does the probabilistic definition of a cause differ from a deterministic definition? Which definition do we use in causal inference?
4. If we know to what degree D causes Y, can we determine the exact value of Y from knowing D?
5. *"How long should I study for the upcoming exam?"* Is this a causal question?
6. What is the difference between an association (or correlation) and causation?
7. The central bank in your country decreased interest rates, and in the following months, the economy, measured as Gross Domestic Product (GDP) per capita, grew. Can we conclude that the lowering of interest rates caused economic growth?
8. For each of the following examples, see if you can come up with an example of factors that could introduce confounding bias.
   a) The effect of flashcards on learning outcomes
   b) The effect of an all-meat diet on muscle mass
   c) The effect of education on income level
   d) The effect of nightly story time on a child's IQ
   e) The effect of yoga classes on mental health

2

The Causal Framework

Before we dive into the basics of causal inference, an important step is to get familiar with some causal terminology. In this chapter, we pick up some new terms. We start with the different words we can use to express causality. Then we explore some different types of causal questions. Finally, we learn about two different approaches to causal analysis: observational studies and experimental studies.

Expressing Causality in Words

While the most obvious way to express a causal relationship between two variables is to say one *causes* the other, causal statements do not have to include the word *cause*. There are other verbs we use to communicate causality, such as *influence, determine, increase, decrease, produce, prevent,* and *enhance*.

While these words are relatively easy to spot in causal statements, there are also myriad other more nuanced causal phrases. For instance, the simple statement "Humphrey broke the window" is a causal statement because he caused the window to break. Or the phrase "Darkness scares me" can be rewritten as "Darkness causes me to scare." Causation is, therefore, a label for a slew of terms. Take a look at the following statements and decide which ones are causal statements.

1. "We find that access to health insurance is a determinant of physical health."
2. "Going to public colleges leads to higher future earnings."
3. "We aimed to evaluate the potential impact of PM2.5 exposure on time to pregnancy."[1]
4. "We find no evidence that hospitalization due to COVID-19 is affected by the obesity rate in a country."
5. "To sleep better, you should not drink alcohol at least a few hours before sleep."

None of the statements above explicitly uses the word *cause*, but they all imply causality. In the first statement, access to health insurance causes physical health. In the second one, going to public colleges causes more future earnings. In the third statement, the researchers try to investigate the effect of PM2.5 exposure on time to pregnancy. In the fourth statement, the claim is about how the obesity rate causes higher hospitalization rates due to COVID-19. In the last statement, however, causation is hidden in the sentence, but the sentence makes it clear that not drinking alcohol affects sleep or causes better sleep.

So all of the statements above infer causality in one way or another. We can also play with words to infer causality. Pay attention to the following sentences.

1. "People who meditate are less likely to be anxious."
2. "If I meditated, I would be less likely to be anxious."

Is the first sentence a causal statement? No. It simply suggests an association between meditation and anxiety. Across the different people in our sample, meditation and anxiety are associated. If you're not paying attention, you might interpret this statement as a

---

1. PM2.5 refers to particulate pollutants that are 2.5 microns or smaller in size.

causal claim, and people will often use language like Statement 1 to get you to believe there is a causal link without having to explicitly lie about there being one. On the other hand, the second statement implies that meditation affects anxiety. It suggests that my anxiety level would go down if I meditate. The grammatical differences are subtle, but the implications are significant. Now, how about these statements?

1. *"Alcohol use was linked to over 74,000 cancer cases last year."*
2. *"Researchers find that Black people are more likely to be arrested by the police."*
3. *"$1,000 spending on housing aid is associated with a decrease in the poverty rate by 1.2 percentage points."*

The authors of these statements are careful not to claim any causality here. For instance, in the first statement, two things being linked together do not say anything about which one causes the other. The second statement implies an association between being black and getting arrested by the police, but it does not say that race causes the likelihood of being arrested to change. Remember, correlation is not causation. Similarly, the last statement uses the words *associated* pointing at an association and not causation between spending on housing aid and the poverty rate.

Basic Causal Inference Terminology

As we discussed, causal analysis is about causes and effects. In statistics, we usually refer to the variable that is the cause as the treatment or exposure variable. We also refer to the variable that is caused by the treatment variable as the outcome variable. Both of these variables are important. If there is a cause, there should be an effect. If we are interested in a cause, we are curious to see the impact of that cause on something, and that "something" is our outcome. So treatment is the cause, and the outcome is the effect.

Note that causes always precede effects. Put another way, the timestamp of the cause should always be before the timestamp of the effect. For instance, throwing the rock happens *before* the breaking of the window. An educational intervention always happens *before* academic outcomes are changed. A vaccine is injected *before* health outcomes are measured.

Let's see what treatment and outcome variables are in some of the examples we have so far discussed in this chapter:

1. *"We find that access to health insurance is a determinant of physical health."* Treatment is access to health insurance, and the outcome is physical health.
2. *"Going to public colleges leads to higher future earnings."* Treatment is going to public colleges, and the outcome is future earnings.
3. *"We aimed to evaluate the potential impact of PM2.5 exposure on time to pregnancy."* Treatment is PM2.5 exposure, and the outcome is time to pregnancy.
4. *"We find no evidence that hospitalization due to COVID-19 is affected by the obesity rate in a country."* Treatment is obesity rate, and the outcome is hospitalization due to COVID-19.
5. *"To sleep better, you should not drink alcohol at least a few hours before sleep."* Treatment is drinking alcohol, and the outcome is sleep (whether quantity or quality of sleep is unclear from the statement).

So you get the idea. We call some of the variables above a *treatment* even if they are not quite treatments in the medical sense. For instance, going to college would not be a treatment a doctor would prescribe, but we call it treatment in the causal inference lingo anyway.

Now that we know what to call our variables, let's talk about how we quantify them. Some treatment and outcome variables are categorical or discrete, while others are continuous. A categorical variable is a variable that is categorized into states or groups. For example, blue/green/red, on/of, or true/false. A discrete variable is a variable that can only take integer values such as 0, 1, 10, or 1,000. A continuous variable can take any one of an infinite set of values on a continuous scale.

Here are some examples. In a study of the effect of marital status (single, married, divorced, etc.) on happiness, the treatment variable is categorical. In a study of the effect of parental income on the choice of college (private or public), the treatment variable is continuous, but the outcome variable is categorical. But because the outcome variable can only take two categories, sometimes people refer to them as binary.[2]

If the treatment variable is categorical, especially if it can take two categories of treatment and no-treatment, the group that receives the treatment is called the *treatment group*, and the group that does not is called the *control group*. The control group usually serves as the basis for comparison. We can also have multiple treatment groups for various treatment options—pill A versus pill B versus pill C.

In a causal study, sometimes treatment is something that can be modified or assigned. For instance, we can give a drug to certain people and not assign it to others. The same is true for enrollment in a training program. The researcher can choose who receives the treatment and who does not.

However, some treatments can only be assigned in theory but not practice, usually due to ethical or practical issues. In a study of the effect of wearing a car seat belt on road casualties, we cannot assign the treatment to some and not to others and see what happens to them. Hypothetically, we can imagine doing that, but it would be unethical to do so in practice. The same goes for smoking; it would be unethical for a researcher to force people to smoke.

---

2. A binary variable usually refers to a categorical variable with only two categories.

The last type of treatment is a treatment that cannot be assigned whatsoever in practice or theory. We cannot even imagine assigning them to people. Think about gender, height, blood type, and so on. In a study of the effect of blood type on vulnerability to a specific disease, we cannot assign different blood types to people in the study.

Now that you know about the different types of treatments, let's go over two more terms, treatment effects and units. The *treatment effect* is the effect of a treatment on the outcome. For instance, if the treatment is giving universal basic income[3] to families, and the outcome is employment status, then the treatment effect is the effect of universal basic income on an individual's employment status. If we average the treatment effect across all individuals in the sample, we call it the *average treatment effect* (sometimes ATE).

We talked about treatment, outcome, and average treatment effect; the treatment effect averaged across all units in the study. What are units? In statistics, a *unit* or a *subject* is a member of a set of entities to be studied. If we are interested in the causal effect of sodium on blood pressure among individuals, then individuals in the study are the units or subjects in our study. If we are interested in the impact of medical costs on population health across countries, then each country in our sample is a unit or subject. You get the idea. Some common examples of units or subjects in causal studies are individuals, cities, zip codes, states, countries, animals, objects, schools, classrooms, and so on. Identifying the subjects in a causal study is usually the first step.

Different Types of Causal Questions

There are generally two forms of causal questions. The first type are causal questions where we are interested in the effect of a specific

---

3. Universal basic income (UBI) programs are government programs designed to reduce poverty and replace other nonmonetary governmental programs such as tax credits or vouchers. Through these programs, citizens receive a set amount of cash on a regular basis.

treatment variable on a particular outcome variable, such as the effect of smoking on lung cancer. Another example would be the effect of parental income on a child's future income. The treatment in both of these studies is specific: smoking in the first study and parental income in the second. These kinds of causal questions are called forward causal questions. In *forward causal questions*, we are mainly interested in the *effect of a cause*.

In contrast, the second type is called *reverse causal questions*, where we do not have a specific treatment in mind. We are mainly interested in why a particular outcome happens. In other words, we are interested in the *causes of an effect*. We can rephrase the two causal questions above to make them reverse causal questions. For the first study, if we asked what causes lung cancer, we have broadened our scope beyond just smoking. We are interested in anything that contributes to lung cancer, such as smoking and genetic factors. In the second question, if we asked what determines a child's future earnings, we would ask a reverse causal question. Again, in this question, we are interested not only in the effect of parental income on future earnings but also in anything else that is important in determining a person's future earnings.

To recap, in forward causal inference questions, we ask, "What is the effect of?" and in reverse causal inference questions, we ask, "What are the causes of?" Both kinds of questions are important. However, forward causal questions are easier to answer empirically simply because the treatment is known. Additionally, most policy-oriented and scientific questions are forward questions; we are interested in the effect of a specific policy or intervention on the outcome.

Let's look at an example. Consider the question from earlier in this chapter: "Does the obesity rate affect hospitalization due to COVID-19?" The outcome variable in the study is hospitalization due to COVID-19. The treatment variable is the obesity rate, and because the treatment variable is specifically stated, this is a forward causal question. How about the units or subjects in the study?

Interventional and Observational Causal Studies

When it comes to causal inference data, we are either using interventional or observational data. You might have heard of randomized experiments, randomized trials, or randomized controlled trials. These terms more or less refer to interventional settings. In interventional settings, the designers of the experiment can intervene in the study and be in charge of treatment assignment or have some control over it. For instance, an interventional setting is a research study where the experiment designer decides who can enroll and who cannot enroll in a program. In interventional settings, treatment assignment is usually random.

On the other hand, in observational settings, there is no intervention, experiment, or randomized assignment. In observational settings, treatment assignment is usually determined by the subjects themselves. Therefore, the researcher cannot intervene in the study and simply *observes* the data. We will see in a future chapter that answering causal questions with observational data is more challenging than with interventional data.

Let's look at an example of each. Imagine we are interested in the effect of wearing masks on contracting the coronavirus. This is a causal question where the treatment variable is wearing a mask—let's say as a binary variable where 1 means the person wears a mask regularly and 0 means the person does not. The outcome variable is whether the person contracts the virus during a specified period.

If we want to answer this question using interventional data, we would potentially select a representative sample of the population in which we are interested. A representative sample is a sample similar to the population we intend to investigate. If we want to study the effect of masks on getting COVID in the United States, we would like a sample that is representative of the US population in terms of demographic and socioeconomic factors. For example, if 15 percent of the US population is above the age of sixty-five, we would like a sample with almost the same share of sixty-five-year-olds.[4]

Once we have a representative sample, we randomly select (let's say by flipping a coin) half of the sample and ask them to wear masks regularly. We then ask the remaining half not to wear a mask regularly. Then we would wait and see which ones get COVID over the specified time. The average effect of wearing masks on getting COVID would be the difference in the share of those who contract the virus between the two groups (treatment and control groups).

If we do not have the luxury of using interventional data to answer the causal question above, we need to look for observational data. In this case, we can survey individuals and ask them if they have been wearing masks regularly and whether they contracted COVID during the specific period we have in mind. Some people wore masks regularly, and some did not. It was not the researcher who determined who should wear a mask and who should not. We will soon see that we cannot simply compare the COVID rates between the two groups: the causal effect estimated this way is potentially biased due to confounding bias.

Chapter 2 Problems

1. Which of the following statements from news articles claim causation and which claim association?
   a) "Music lessons improve kids' brain development."
   b) "Housework cuts breast cancer risk."
   c) "The luckiest people are 'born in summer.'

---

[4]. An issue with answering this causal question with a randomized experiment is noncompliance. Here, subjects of the study may not necessarily follow the assignment to wear or not wear masks. We return to the issue of noncompliance in randomized experiments in chapter 4.

d) "Trust in government linked to work attitudes!"
e) "Later school start times reduce car crashes"
f) "Baby teeth may identify ADHD!"
g) "Checking phones in lectures can cost students half a grade in exams."
h) "Dogs owned by men are more aggressive!"
i) "Sincere smiling promotes longevity."
j) "Deep-voiced men have more kids!"

2. For the causal statements among the statements above, what is the treatment, and what is the outcome variable?
3. Consider the following causal questions. For which ones is it possible to assign the treatment? If it is, do you think it would be practical/ethical to do so?
   a) What is the impact of wearing helmets on cycling accidents?
   b) Are women athletes paid less?
   c) Does providing free milk at schools affect academic outcomes?
   d) What is the impact of daily school recess on classroom behavior?
   e) Does the socioeconomic status of parents during childhood influence the probability of going to college?
   f) Does cell phone use disrupt sleep?
4. In the causal questions above, what would be our units or subjects?
5. Go through the following claims from media articles and determine whether they are causal or correlational. How would you rephrase the statement if you wanted to switch the type of relationship (causal instead of correlational or correlational instead of causal)?
   a) "Social support improves the mental well-being of older adults."
   b) "Child anxiety is linked to drug use."

c) "Having a fan in the room seems to cut infants' risk of crib death."
   d) "Tooth loss in the elderly linked to mental impairment!"
   e) "Obese high school students are less likely to attend college, research shows."
6. From the statements above, pick the ones that are causal and determine the outcome and the treatment variable in the study. How could these variables be measured, and would they be discrete or continuous?

3

Causal Graphs and Causal Paths

Expressing Causality Using Causal Graphs

Just like expressing causal claims using words, we can describe them visually. Take the causal statement mentioned earlier: "We find that access to health insurance is a determinant of physical health." In this statement, the treatment variable is access to health insurance, and the outcome variable is physical health. As we saw, we can briefly show the causal relationship using a causal graph with two nodes; one for the treatment variable and one for the outcome variable. To show that one of these variables is the treatment and the other is the outcome variable, we use a darker color for the outcome variable. For more clarity, we keep this protocol for the rest of the book.

This causal graph includes two nodes or variables and a link or edge between them. The link goes from the left node to the right node, indicating that the left node—access to health insurance—affects the right node—physical health—and not the other way around.

Now, consider a more complicated example with more than two variables. In this causal graph, access to health insurance is still a determinant of physical health, but we also have two new variables: income and nutrition. Income influences nutrition, which in turn affects physical health. We can show this with an edge pointing from income to nutrition and another edge pointing from nutrition directly to physical health. We also have an edge pointing from income to health insurance. This represents the fact that income is a determinant (or cause) of access to health insurance. This causal graph, containing four nodes, is quite a bit more complicated than our first one, but that's okay. In fact, the whole reason we use causal graphs is to map out more complicated causal patterns, such as this one.

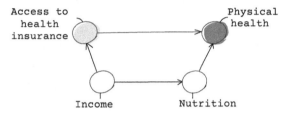

The two nodes reflecting income and nutrition are shown in white to distinguish them from treatment and outcome nodes. Next we go over some of the main elements of causal graphs.

Elements of Causal Graphs

Causal graphs are also called directed acyclic graphs (DAGs). A causal graph consists of nodes and directed links. We say directed link because the direction tells us about the direction of causality between nodes. A lack of direction indicates no causality and merely an association between two nodes, such as the one below.

A path is a combination of links from a starting node to an end node. For instance, in the causal graph below, there are two paths between X1 and Y.

- X1 → X2 → Y
- X1 → X2 → X3 ← Y

Note that the direction of the links in a path can change (X2 → X3 then X3 ← Y).

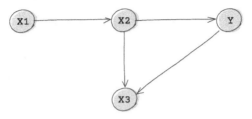

Causal graphs are called *directed* due to directions on the links. We also call them *acyclic* because there are no closed paths—that is, there should not be any path that starts from a node and ends in that node without changing direction. The following graph is not a causal graph because there is a cycle in it.[1]

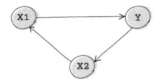

Causal graphs are helpful in communicating a causal problem. In a causal graph, nodes tell us about the relevant variables in a study, and edges or arrows tell us about how nodes cause each other. Moreover, if there is no arrow between two specific nodes, we assume there is no direct causal link between those two nodes.

---

1. In practice, we do not have to worry about cycles in causal graphs because of what we briefly discussed in chapter 2: causes always precede effects. One could argue that if A causes B and B causes A, then we have a cycle and the associated graph is not acyclic. However, causes always happen at a time before effects are measured. So A causes a future version of B, and B causes a future version of A that is no longer the same as A.

What Is a Direct Causal Path?

Causal graphs are not just visual representations of causal relationships among variables. They can also help us identify biases in causal estimates. For this purpose, we first need to distinguish causal and noncausal paths between the treatment and outcome nodes.

Any path that goes outward from the treatment variable and leads to the outcome variable is a causal path. Other paths between the treatment and outcome that do not fit this description are called noncausal paths. We will see that causal paths are usually the ones in which we are interested.

Here is an example of a causal path: D is the treatment variable, and Y is the outcome variable. The only causal path is D → Y because it goes outward from the treatment node and leads to the outcome node.

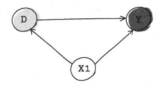

The D ← X1 → Y path is also between D and Y, but note that the path does not point outward from the treatment variable D. Instead, it is directed *toward* the treatment variable. We can therefore conclude that this is not a causal path.

Another example might be helpful. In the next causal graph, variable D is the treatment variable, and variable Y is the outcome variable. Let's first go over all the paths between D and Y.

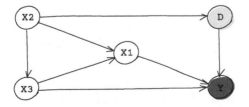

These are the paths:

- D → Y
- D ← X2 → X3 → Y
- D ← X2 → X3 → X1 → Y
- D ← X2 → X1 → Y
- D ← X2 → X1 ← X3 → Y

Some of these paths are causal, and some are noncausal. The only path that is causal is the path that is outward from the treatment variable D. Therefore, the only causal path is the first path. All the other paths are noncausal.

Lastly, let's look at a causal graph with multiple causal paths. Imagine we are interested in the effect of family socioeconomic status (SES) on individuals' income. Therefore, SES is our treatment variable, and income is our outcome variable. For the sake of our example, imagine the following causal graph represents the causal relationships.

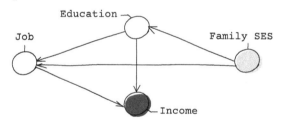

Can you spot the causal paths?

All paths between family SES and Income are causal because they are outward from the treatment variable. We do not have any noncausal paths between the treatment and outcome. These causal paths are

- Family SES → Education → Job → Income
- Family SES → Education → Income
- Family SES → Job → Income
- Family SES → Job ← Education → Income

Screen Time and Obesity in Youth

Smartphones, tablets, and computers have changed our lives. People now communicate, learn, and participate in social and political activities on their devices. But the question we want to ask is are smartphones and computers to blame for the increase in obesity among children? A study used data from more than eleven thousand children in the United States and found that children who spend more time on their smartphones or computers are more likely to suffer from binge-eating behavior and, therefore, be obese.[2] Although the study does not explicitly claim a causal relationship, the wording strongly implies causality to readers. The researchers even caution parents about "the potential risks associated with excessive screen time."

We can summarize the causal claim in this causal graph:

The causal dynamic is more complicated than this causal graph shows. We know technologies such as smartphones have changed our lifestyles, but we cannot take seriously a study based on simple association. We need a more comprehensive picture of the problem. Let's assume we know the following:

- Screen time decreases the amount of time youth spend on physical activity (this is trivial).
- Lower physical activity increases the chances of obesity (this is trivial too).
- Parental education affects the amount of screen time (let's assume previous research has proven this).

---

2. Nagata, J. M., Iyer, P., Chu, J., Baker, F. C., Pettee Gabriel, K., Garber, A. K., ... Ganson, K. T. (2021). "Contemporary Screen Time Modalities among Children 9–10 Years Old and Binge-Eating Disorders at One-Year Follow-Up: A Prospective Cohort Study." *International Journal of Eating Disorders*, 54(5), 887–892.

- Parental education affects children's nutrition (while trivial, let's assume previous research suggests this).
- Nutrition affects obesity (let's say this is medically proven).

If we employ all the assumptions above, we come up with the following causal graph:

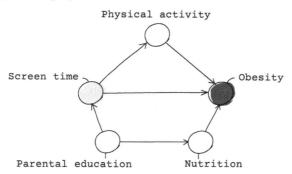

Note that the treatment variable is screen time, and the outcome variable is obesity. Aside from the direct causal path between screen time and obesity, from this causal graph, screen time and obesity are related to each other directly but also through physical activity, parental education, and nutrition. These are the three paths:

- Screen time → Obesity
- Screen time → Physical activity → Obesity
- Screen time ← Parental education → Nutrition → Obesity

Which of the three paths above are causal, and which are noncausal? If we go back to the definition of a causal path, we know that the first two paths are causal paths (one is direct and one is indirect), and the last path is noncausal.

Why is it important to distinguish between causal and noncausal paths? When we look at the correlation between the treatment and the outcome variables, this correlation captures all causal and noncausal paths between the two variables. For us to extract only the causal effect, we need to only look at the causal paths between treatment and outcome.

If we empirically find the correlation between screen time and obesity, we are looking at all links between the two variables. The correlation that we find shows how screen time is related to obesity through physical activity (the causal path) and how screen time is related to obesity through parental education and nutrition (the noncausal path). If we find the overall magnitude of only the causal paths, we have figured out the causal effect of screen time on obesity.

You may ask how we can only extract the effect through the causal paths. To find the *pure* causal effect, we need only to measure the strength of the causal paths, and the only way to do so is to somehow disconnect the noncausal paths like in the following figure. If we block all noncausal paths and we still see a correlation between the treatment and the outcome, then we can attribute this correlation to the causal effect of treatment on the outcome. An important part of a causal analysis is to identify noncausal paths and block them. In chapter 5, we learn how we can block or disconnect a path.

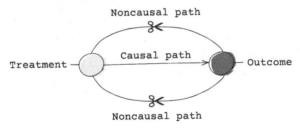

Does Ambient Lighting at Night Cause Myopia in Children?

Myopia, or nearsightedness, is an eye disorder in which light focuses in front of the retina instead of on it. It is a genetic disease, but environmental factors can also impact it.

In 1999, a study by Quinn et al. found a strong association between myopia and night-time ambient light exposure during sleep among children.[3] It was, however, unclear whether this association was causal or not. In other words, it was unclear whether it reflected the causal effect of ambient lighting on myopia among children or just a simple correlation between the two. We might be able to clear up this confusion using causal graphs.

Assuming there are no noncausal paths, the simplest causal graph looks like this.

Is this causal graph sufficient or helpful? Neither. It leaves out important details that are relevant to our causal question. For instance, it turns out that parents who have myopia themselves use night lights in their child's bedroom. Therefore, we can rephrase this as parents having myopia affects having a night light in the child's room. If myopia is genetic, then parents having myopia affects the chances of developing myopia among their children. So parents having myopia also affects the child's myopia. The result is a complete causal graph like the one shown below.

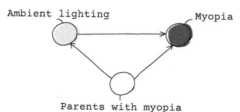

This more complete causal graph shows two paths between the treatment (ambient light) and the outcome (myopia in the child).

---

3. Quinn, G. E., Shin, C. H., Maguire, M. G., and Stone, R. A. (1999). "Myopia and Ambient Lighting at Night." *Nature*, 399(6732), 113–114.

- The ambient light → child's myopia is a direct causal path.
- The ambient light ← parents' myopia → child's myopia is a noncausal path.

As we know, we should only be interested in the causal path. A simple correlation between ambient lighting and a child's myopia, the one claimed in the paper, reflects both paths and does not correctly establish the causal relationship between the treatment and outcome variables. This estimate is said to be biased, and the bias is due to *confounding bias* as we will later see. We should do something about this confounding bias if we are only interested in the causal path. As we learn later, we should block the noncausal path.

But before investigating studies with potentially noncausal paths between treatment and outcome variables, let's look at interventional causal studies that are generally assumed to be free of noncausal paths and confounding bias. In the next chapter, we go over interventional studies.

Chapter 3 Problems

1. In the following DAG, node D is the treatment, and node Y is the outcome. Why is the DAG not valid?

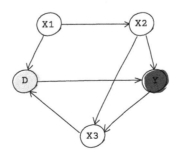

2. Draw causal graphs for the following scenarios:
   a) In a causal study of the effect of providing free meals at schools on students' academic outcomes, prior research suggests the following: Parents' socioeconomic status (SES) affects the amount of time spent with children at home, which in turn affects the child's academics. Parents' SES also affects the neighborhood the child lives in and, consequently, the probability of going to a school that provides free meals (schools in poorer neighborhoods are more likely to provide free meals at school).
   b) We are interested in the effect of the amount of exercise on body mass index (BMI). We know that having a higher income means better access to health care and doctors, which in turn means lower BMI. Income also affects access to gyms, which in turn affects the amount of exercise. Income also affects BMI through diet; a higher income means a healthier diet, and we know diet affects BMI.
3. Describe the following scenario based on the causal graph below. Note that the study is about the effect of playing video games on stress levels among youth.

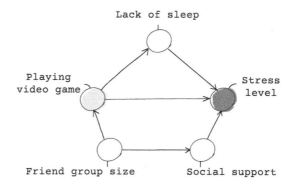

4. Consider the following causal graph. Here D is the treatment and Y is the outcome. List all the paths between the treatment and the outcome. Which of these paths are causal?

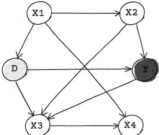

4

Causal Inference Using Interventional Data

So far, we have learned how to frame causal questions and use causal graphs to express causal relationships. We learned that a causal estimate is unbiased when all noncausal paths between treatment and outcome are blocked, and we looked at some hypothetical studies.

In this chapter, we go over a specific type of causal study called an interventional study. These studies are also called randomized experiments. Randomization plays an important role in two key phases of an interventional study: sampling and assignment. Random sampling (or selection) describes how you should select subjects for your experiment from the broader population of interest. Ideally, you aim for a sampling process where each individual in the population has an equal chance of being selected for the study. Random sampling is likely something you have already heard or learned about in a statistics course, so in this chapter we focus on the role randomness plays in the assignment of subjects into treatment and control groups.

In a randomized experiment, assignment to treatment or control groups is also random. We make sure that the assignment is random by using a random number generator or a mechanism like a coin toss to determine which subjects in our study receive treatment and which do not. For example, suppose you have one hundred units (or subjects) in your study. For each unit, you could simply toss a coin. If the coin lands heads, you assign the unit to the treatment group, and if it lands tails, you assign the unit to the control group. If the coin is fair, the coin toss guarantees random assignment and ensures that each unit has an equal chance of being in either group.

If we have multiple treatment groups, let's say we have treatment 1, treatment 2, and control (no treatment), then we use a random number generator to assign individuals in the sample. Number 1 assigns the subject to treatment group A, number 2 assigns the subject to treatment group B, and number 3 assigns the subject to the control group.

Why Is Randomization Important?

Why is randomization so important? We all run our own mini-experiments every once in a while. Even if these informal experiments may not be perfect, they can give us some insights into how we should and shouldn't run causal experiments.

Imagine you haven't been sleeping well lately and you want to know if your alcohol consumption has been interfering with your sleep. We've all had situations like this where we want to know whether a particular habit we have is responsible for positive or negative impacts in our day-to-day lives. You decide to run a little experiment. You drink alcohol one night and see how you sleep. On that night, you are in the treatment group. Another night, you imagine yourself in the control group and see how you sleep without drinking alcohol. Even though there is only one person in your little study, you are hopeful that your investigation reveals something about the effect of alcohol on your sleep.

This mini-experiment, however, is far from perfect. Beyond the small sample size, there is another glaring issue. There are likely important differences you've overlooked between the night you drank and the night you didn't. What if, on the night you had alcohol, you also consumed more caffeine? If that is the case, you should probably question your findings. You have no way of knowing whether any differences in sleep quality between the two nights is a result of the alcohol or the caffeine. Even if both factors impacted the way you slept, you have no way of knowing how much of your sleep deprivation to attribute to the caffeine versus the alcohol.

Randomization and a large sample size can help you fix this problem. Instead of conducting the experiment over just two nights of your choosing, you should instead repeat your experiment over a longer stretch of time, say fourteen nights. You can then randomly assign half of your nights as "drinking nights" and half as "nondrinking nights." If you do this and keep a diary of how you slept each night, you will get a more reliable result. If you want to be even more rigorous, you can extend the experiment over an even longer period of time, with each night counting as one observation in your mini-randomized experiment. Or better yet, you can include a dozen of your friends in your little study.

Note that random assignment of nights to drinking nights and nondrinking nights is key. But why? How exactly does repeating the experiment over many nights and the random assignment improve your experiment? The answer is that random assignment helps balance out all of the other factors impacting your sleep so that their effects, on average, are no different between your treatment nights and your control nights.

Think back to your caffeine intake. In the two-night version of the experiment, you worry that on one of the nights, your caffeine consumption was substantially higher than on the other. This difference throws off the results of your experiment, but by increasing the number of nights and by randomly assigning those nights to treatment and control, you can gain confidence that, on average, your caffeine intake on the drinking nights will be similar to your caffeine intake on the nondrinking nights.

Why not just be mindful of how much caffeine you consume throughout the course of the experiment? That would resolve the caffeine issue for sure (assuming you pay close attention to how much caffeine you're having), but remember, many, many factors can influence your sleep, and you may not even be aware of many of them. Paying attention to caffeine in addition to alcohol may be doable, but you would also have to meticulously monitor things like exercise, exposure to sunlight, stress levels, diet, naps, and many others. Carefully controlling all of these outside influences is impossible, whereas randomization balances out these effects, even for factors you are unaware of.

To generalize this a bit more, in a causal study, you're interested in the effect of a particular treatment on a particular outcome variable. The outcome variable, however, may have many other variables it is listening to. Experimental studies help us cancel out the noise from all these other factors through random assignment. In this way, randomization allows you to isolate the particular causal relationship that interests you.

Once we've drowned out the noise from all of these other factors, we are in a much better position to compare differences in the outcome variable with and without treatment. Making this clean comparison is what causal inference is all about. For every forward causal question, you are interested in how the outcome variable would compare under different treatment regimes. When you have other variables influencing the outcome, your comparison becomes tainted or biased. The key to an *unbiased causal estimate*, a term statisticians use to say that the causal estimate is as close as possible to its actual value, is in having comparison groups that are more or less similar. In an ideal setting, we would observe our sleep in two otherwise identical worlds. In one of these worlds, we would introduce a single change: we drink at night. In the other, we wouldn't. If we could run this experiment, we would have a perfect measure of the causal effect of drinking at night, but such an experiment would only be possible if we could time travel or access the multiverse. Since we can't do either of those things, our next best option is to randomize. We randomly assign ourselves to drink on some nights as opposed to others and compare our sleep. We can choose a large number of nights for our experiment to be meaningful. By randomizing and increasing the overall number of nights in your study, you are making sure that your comparison of drinking and nondrinking nights is fair.

Let's turn to a different example. Suppose we are interested in the effect of smoking on lung cancer. Our goal is to compare the outcome variable, lung cancer, under two different treatment regimes: smoking and not smoking. To make this comparison, we want the group of smokers to be as similar as possible to nonsmokers, other than the one difference we want to isolate, which is that one group smokes and the other does not. More specifically, we want the group of smokers to be similar to the nonsmokers in ways that matter to the causal question. If we make sure that the two groups are similar, then we guarantee any difference in their rate of lung cancer is due to smoking and not other factors such as age.

Let's say age is important in studying the effect of smoking on lung cancer. If so, we want the groups of smokers and nonsmokers to be very similar regarding age. If 20 percent of smokers are young, we also want 20 percent of nonsmokers to be young. If 15 percent of smokers are fifty or older, we want the same ratio among non-smokers. The more we are able to ensure that the treatment group (smokers) and the control group (nonsmokers) are similar, the more reliable our comparison will be.

A Numerical Example

Now let's turn to a numerical example. Employer-sponsored wellness programs usually offer gym access, weight loss programs, and mental health support for employees. Imagine you are an employer and are interested to see if the wellness program you started at your company works. More specifically, you want to know if providing the wellness program reduces your company's healthcare costs.

Let's assume for a minute that you can force some employees to use your company's wellness services and stop others from using them. You justify your harsh approach to study design by insisting that everybody is relatively randomized to the groups and there is no favoritism. To answer the causal question, you run the program for one full year and, in the end, observe the health outcomes of those who actively participated in the program to those who did not.

The treatment is active participation in the program, and the outcome is your annual contribution to your employees' health expenditures. The subjects in your analysis are the employees.

Let's look at some numbers. You have two hundred employees in your program, one hundred of whom were randomly assigned to the wellness program and one hundred of whom were prevented from accessing it. The one hundred who are denied access are the control group.

The table below shows some information about the treatment and control groups. Under randomization, the two groups are more or less similar. By randomly dividing employees into treatment and control, we ensure that the two groups are similar on average.

TABLE 4.1 Variable statistics for treatment and control groups

Group	% female	% age 50+	% w/ preexisting conditions	Avg. costs
Treatment	42%	51%	19%	$10,000
Control	39%	50%	20%	$15,000

But if the two groups are supposed to be similar, why are their healthcare expenses—the outcome variable—different? The two groups should be similar, but we know one group had access to the wellness program while the other did not. As a result, we see that their healthcare expenses are different by $5,000 on average. We know the only reason the two average healthcare expenditures differ is the wellness program. So we can call this the effect of the treatment in the study—or simply the treatment effect. We succeeded in our task. We are happy because the wellness program seems to be working and making a difference.

Understanding Treatment Effect

Let's dissect the treatment effect we found in the previous example. What does it actually represent? An important question to ask is: Does the $5,000 treatment effect apply to everyone in the study? In other words, did each person in the treatment group have exactly $5,000 less in healthcare expenses because of the program?

The answer to the question is no. The number we are talking about is an average. First, randomization does not guarantee that for each person in the treatment group, we have a similar person in the control group. In other words, we do not create doppelgangers at an individual level. Because of randomization, the treatment and the control groups are similar as a whole, or on average.

Second, we compared their average expenditures to find the treatment effect. We subtracted the average costs in the control group, $15,000, from the average costs in the treatment group, $10,000. Therefore, the treatment effect, -$5,000, is called the average treatment effect. More specifically, the effect of participating in the wellness program is, on average, a reduction in healthcare expenditures by $5,000. For some individuals in the sample, the treatment effect might be larger than $5,000, and for some it might be smaller. We can assume that across the sample, the treatment effect averages out to $5,000.

Explaining Randomized Experiments with Causal Graphs

We can look at how randomized experiments usually give us unbiased estimates of the treatment effect by looking at causal graphs and causal paths.

In estimating causal effects, as we saw in the previous chapter, we should keep the causal path in mind and block all noncausal paths. But what are the paths in a randomized experiment? Imagine we are interested in the effect of taking melatonin on hours of sleep, and we use a randomized experiment to find our answer. We give actual melatonin pills to a randomly selected subset of our sample and give placebo pills to the other half. The causal path that we are interested in is the direct path from the treatment variable, taking melatonin, to the outcome variable, hours of sleep. Below is the causal graph that shows this direct causal path.

Note that because we used a randomized setting, the only factor that affects the treatment variable is a coin toss. Nothing else determines who receives the treatment and who does not. So if we want to add the coin toss node to the causal graph, we have something like the graph below.

You might argue that this is a very simplified causal graph, and other factors might be at play. Let's think about some of them. For instance, many other factors affect hours of sleep, like exercise, alcohol, and caffeine. We can add them to the causal graph.

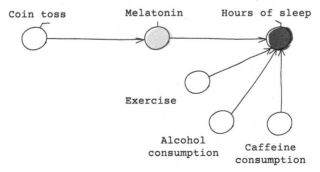

Do these factors only affect the outcome and not the treatment? Definitely! Remember, the coin toss is the only factor that determines—read "affects"—the treatment. Since individuals in our study were randomly assigned to treatment and control, none of their other characteristics, such as age, gender, or caffeine intake, affected their treatment versus control status. That's why the graph makes sense – it reflects how only the coin toss determined who got the treatment.

Now, we know that for the treatment effect in the causal graph above to be unbiased, it should only capture the causal path between the treatment and the outcome. This means that all the noncausal paths should be disconnected. Are there any noncausal paths above?

If you refer back to the definition of noncausal paths, they are paths between treatment and outcome that are inward to the treatment. Are there such paths? No. None of the paths in this causal graph point inward into the treatment. All those arrows going to the outcome variables do not eventually lead to treatment, so they do not form any paths. The only path between treatment and control is a causal path, which is the direct path between the two variables. Note that this is the path in which we are interested.

Although this was a specific example, randomized experiments will generally always create a situation with no noncausal path between treatment and outcome. As a result, you can find the treatment effect by simply finding the association between the treatment variable and the outcome variable. The causal estimate that we find through a randomized experiment is unbiased.

A Little History of Randomized Experiments

One might guess that randomized experiments were only used in the modern era. Avicenna, the Persian polymath and one of the world's greatest physicians, laid out some of the principles of doing randomized experiments in medicine in AD 1025.

However, these experiments were mentioned in the book of Daniel in the Bible much earlier. The story goes that roughly around 562 BC, King Nebuchadnezzar, the king of Babylon, wanted to see the benefits of a vegetarian diet. To find his answer, he ordered some of his people to only eat meat and drink wine. Then he told others to eat only vegetables, including legumes and water. The experiments continued for ten days, after which the king realized that those on the vegetarian diet appeared to be better nourished than the rest.

Perhaps one of the earliest scientific use cases of randomized experiments was done by James Lind, the Scottish naval doctor, during one of his voyages in 1747. He used an approach similar to modern clinical trials to find a cure for scurvy when nobody knew much about it and its causes. If you do not know about scurvy, Lind's experiment will reveal what causes the disease. As Lind described it, many sailors showed signs of bad gums, spots, and lassitude, with weak knees during the voyage. Being on the ship, he did not have many tools at his disposal, so he started dividing patients into groups to ensure that the groups were as similar as possible. He then administered different remedies to each group ranging from vinegar, water gruel sweetened with sugar, fresh mutton broth, and pudding to oranges and lemon. After monitoring the patients, Lind noticed that those who had received lemon and oranges (hence, vitamin C) saw the most considerable improvements in the symptoms to the point that only six days after the treatment, they were back on their feet and working. His experiment revealed that scurvy is caused by vitamin C deficiency.

Since then, randomized experiments have been a key in understanding medical sciences. The first known randomized experiment in psychology dates back to 1884, when psychologist Charles Pierce wanted to understand people's perceptions of others' weight. Harold Gosnell did the first randomized experiment in political science in 1924. Ronald Fisher performed the first in agriculture in 1927, and Edward Chamberlin performed the first in economics on demand and supply curves in 1948.

By the 1960s, randomized experiments were being used in policy evaluations. One of the early cases of large randomized experiments in policy was in Mexico around the social program PROGRESA, designed to help the poor. The government rolled out the program in stages, creating *natural* control and treatment groups. One of the ways of doing randomized experiments in policy evaluation when it is hard to deny people the treatment is to roll out the program in stages.[1]

Randomized experiments have been one of the most important ways to learn about our world and societies. Without randomized experiments, some medicines would have never become accessible to the public, and many effective public policies would have gone unnoticed. Even though randomized experiments were primarily used in clinical studies, they are increasingly used in hard sciences, social sciences, and policy design.

Shortcomings of Randomized Experiments

We saw that we could exploit experimental or observational data to answer causal questions. In the next chapter, we learn about observational studies and their challenges, but you might be struggling with a question. If using randomized experiments makes answering causal questions so easy, why do we not use them all the time? Why are they not used more often in medicine and social sciences?

In short, randomized experiments are not always feasible. Let's explore some of the reasons why.

Ethical concerns: Randomizing subjects into treatment and control can sometimes be unethical. Think about a study about the effect of smoking on lung cancer. Can you imagine forcing some participants to smoke just because the flip of a coin assigned them to a treatment group? How about a study on breastfeeding and a child's

---

1. Note that rolling out a policy in stages doesn't necessarily create a randomized setting because many things may change over time. We can only stage the policy for finding the treatment effect in cases when we can't deny the population the treatment.

health? These are cases where either there are unintended adverse consequences for those in the treated group, or it is unethical to deprive the control group of the treatment.

Blindness: In an ideal randomized experiment, participants should not know whether they are in treatment or control groups. Not knowing which group they are in prevents subjects from changing their behaviors in ways that might interfere with the results of the study. This is why in medical science, researchers usually use placebos—fake pills and treatments—to hide treatment status. Subjects do not know whether the pill they are taking is the actual treatment or not. This is called blinding subjects to their treatment status. In policy evaluation and social sciences, blinding is often not feasible. For instance, imagine you're trying to evaluate the effect of a universal basic income. You can create a treatment and a control group, but you can't prevent your subjects from knowing what group they are in. Your subjects will either receive added income as part of the treatment or they won't.

Or suppose you are interested in using a randomized experiment to see whether screen time affects obesity. You cannot create a fake version of screen time. Subjects either spend time in front of their TVs or computers or they do not.

How is lack of blindness problematic? If subjects know that they are participating in a study about screen time and obesity and clearly see that they are in the treatment group, they may change their behavior. They may get stressed out about being in the treatment group. Or they may want to compensate for higher screen time and exercise more. In either of those cases, the difference in the outcomes of the two groups can no longer be solely attributed to the treatment—screen time—and could, instead, be the result of other behavioral changes.

Issues regarding informed consent: In the above example, you may ask why we have to let subjects know about what kind of study they are in. The less they know about the study, the fewer undesired changes in their behaviors. Most randomized experiments need to be approved by a review board to make sure they do not cause any harm to human or animal subjects. Additionally, the review boards usually ask researchers to inform the subjects about the investigated treatment and get their consent before the study. We refer to this as informed consent. This policy was introduced after some researchers took their experiments too far and exposed subjects to physical and mental risks.

What could be a negative consequence of informed consent? If you have to let your study participants know everything about what you are interested in and what you are measuring as an outcome, it may lead to changes in their behaviors, regardless of whether they know they are in the treatment group or not. Imagine running a randomized experiment where you want to see whether women are more likely to return money found on the street than men. In that case, you have to let your subjects know that this is a study about their altruism. The challenge is that if the subjects are fully aware of the intent of the study, subjects may change their behavior.

Noncompliance: Another issue with randomized experiments, related to the fact that researchers cannot always force participants to take the treatment, is noncompliance. Noncompliance arises when we assign subjects to treatment or control, but they do not stick to the assignment. Some of the treatment group subjects do not take the treatment, and some of the control group subjects somehow get their hands on the treatment offered. As a result, even if the treatment assignment is random, the treatment received may not be. For instance, imagine in a clinical study, we randomly assign a supplement pill for joints, but not all those in the treatment group take the treatment. For instance, we realize that those who already have some joint issues are more likely to take the treatment. As a result, treatment received is no longer random and is related to some individual traits, in this case, preexisting conditions.[2]

Attrition: Another issue with randomized experiments is attrition. Attrition happens when subjects voluntarily or involuntarily

---

2. Note that if non-compliance is random, i.e., it does not correlate with personal traits, it is negligible.

drop out of the study. Attrition is usually an issue with longer-term randomized experiments where we monitor subjects over a long period of time. Suppose we are studying the long-term effect of a childcare program on parents' labor market outcomes. We may notice that six months into the study, some of the subjects, especially those who are busier and have little incentive to continue being in the study, drop out. Again, you can think of attrition, causing treatment received to no longer be random. So the type of problem that attrition causes is similar to noncompliance.

Not everything can be randomized: A lot of the phenomena that we like to study—for instance, those in macroeconomics—cannot be randomized. A cross-country causal study of antipoverty programs on inequality cannot be randomized because we cannot randomly assign policies to different countries. Neither can a study on the effect of inflation on unemployment. Sometimes we can use randomization in theory, but it will be costly in practice. For instance, a study on the long-term impact of universal basic income on wellness is expensive to run as we need to give a relatively large group of individuals some basic income for the desired period to see the impacts.

Spillover between the units: To capture an unbiased treatment effect, we need to find the true causal effect of the treatment for each unit in a randomized experiment. This requires that each participant's outcome be influenced solely by their own treatment and not by the treatments received by others. However, it is not always possible to prevent the interference of the units with each other.

To understand this, let's go over a short example. Consider a randomized experiment aimed at assessing the effectiveness of a new educational program in improving student performance. We choose a few schools to recruit subjects for our study. Students in these schools are randomly assigned to either participate in the new program (treatment group) or continue with the standard curriculum (control group).

While the design seems straightforward, satisfying the no-interference assumption can be challenging in this context. This is because students often interact with each other, sharing insights, teaching methods, and study materials. If a student in the treatment group shares key aspects of the new program with a friend in the control group, the outcome of the control group student could be influenced by the treatment. This kind of peer interaction violates the no-interference assumption, as the outcomes of control group students are no longer independent of the treatment assigned to their peers. Consequently, the estimated effect of the educational program may be biased, complicating the interpretation of the experiment's results. This is often an issue in settings where social interactions are prevalent.

Given the shortcomings mentioned earlier, randomized experiments are not always feasible, and researchers sometimes have to exploit observational studies to find answers to critical causal questions. In the next chapter, we will see how we can answer causal questions when we do not have the privilege of conducting a random experiment.

Chapter 4 Problems

1. Imagine we want to conduct an interventional study to understand the effect of government-provided free childcare on mothers' employment. In this setup, the treatment group receives free childcare, while the control group does not. Our goal is to assign half of the participants to the treatment group using randomization. To explore how randomization helps create similar treatment and control groups, we'll simulate the study. Suppose we're particularly concerned about whether a relevant variable—mother's education—is balanced between the two groups. For simplicity, we'll treat

education as a binary variable: either the mother has a college degree or she does not. We know from population data that about one-third of mothers have a college degree. To simulate the randomized experiment, use two different colored dice. The first die represents treatment assignment: if it lands on 1–3, the mother is assigned to the treatment group; if it lands on 4–6, she goes to the control group. The second die represents education level: if it lands on 1 or 2, the mother has a college degree; if it lands on 3–6, she does not—mirroring the one-third probability. Try this simulation with samples of 20, 50, and 100 participants. For each run, record how many participants in the treatment and control groups have a college degree. What would you expect to observe about the balance of the education variable between the two groups?

2. The simulation above only considered one potential confounding variable, but the idea is that in a randomized experiment, all relevant and nonrelevant variables should be balanced. We would like to investigate if another variable capturing mothers' eye colors will end up being balanced across the treatment and control groups. Note that this may not be a relevant variable in our study, but regardless, due to randomization, it should end up being balanced. Imagine the eye color distribution follows the distribution in the United States based on the following numbers: 40 percent brown, 30 percent blue, 20 percent hazel, and 10 percent green. How would you change the simulation above to check if the control and treatment groups are similar?

3. You just read about a study investigating the effectiveness of a data literacy job training program on future earnings. The program is accessible to anyone and is free, but it takes six months to complete. The study finds the average treatment effect of the training program is an average increase

in earnings of $2,000 per year. A friend asks you, an expert in causal inference, whether you recommend participating in the job training program. Which of the following would be your advice?

a) I would not recommend the program. The study says nothing about the effect that the training program would have on my friend.

b) I would recommend the program on the grounds that it will increase my friend's earnings by $2,000 a year.

c) I would advise my friend to weigh her own costs and benefits of enrolling in the program. I would emphasize that, on average, the training program increases earnings by $2,000 per year, but this means any individual's benefit could be higher or lower than $2,000 per year.

4. Consider the DAG from the practice question in chapter 3. The question was about the effect of exercise on BMI. How would the DAG look like had the assignment to treatment (assignment to exercise) been randomized? As a reminder, the DAG originally looked like this:

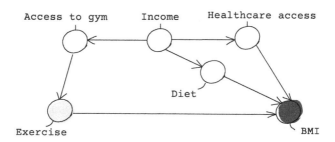

5. The DAG below shows the graph for a randomized experiment. The treatment assignment is determined by a coin flip. What are the causal and noncausal paths?

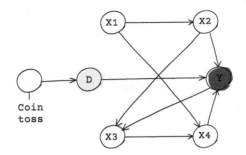

6. Consider you want to study the following causal relationships. What factors might make it difficult to conduct an interventional study?
   a) The effect of race on wages
   b) The effect of using seat belts on traffic casualties
   c) The effect of female elementary school teachers on future earnings
   d) The effect of drinking coffee on productivity
   e) The effect of stress on kindness

5

Causal Inference Using Observational Data

We saw that in a randomized experiment, there is only one path between the treatment and outcome, and that path is the causal path in which we are interested. In the case of employer-sponsored wellness programs and healthcare costs, we saw that if we randomized subjects into treatment and control, the simple causal graph that explains the problem is something like this:

As you can see, no node other than the coin toss leads to the treatment variable since the treatment assignment is randomized. You can also see that the only path between the treatment and outcome nodes is the causal path in which we are interested. So there is nothing to worry about here.

Imagine that we do not have the means to perform a randomized experiment to understand the causal question above. You are given a dataset of a company's employees who did or did not participate in a wellness program. These employees voluntarily decided to participate in the program. You also have data on their healthcare costs one year after the wellness program started.

This example is a case of an observational study. In observational studies, things are a bit different. We previously defined observational studies as those in which the researcher makes no attempt to influence the treatment assignment or the outcome and the research team is only observant. In these studies, subjects self-select into treatment or control groups. What is so special about this self-selection?

Confounding Bias in Observational Studies

In the observational study above, other factors likely belong to the causal graph. One obvious variable is being healthy. It needs no justification to assume that healthier employees are more likely to participate in the wellness program. Healthy employees also have lower healthcare costs. Showing this in a causal graph is simple.

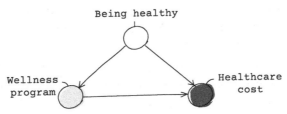

We now make things a bit more complicated. One of your colleagues tells you that education is also an important factor. She convinces you that more educated employees are more likely to participate in the program. Additionally, education affects healthcare costs through nutrition—that is, education affects nutrition, and nutrition affects healthcare costs. Those who eat healthier get ill less. Let's put all these in the causal graph.

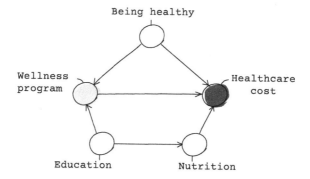

How many paths are between the treatment and the outcome now? We can see that there are now three paths:

- Program participation → Healthcare costs
- Program participation ← Being healthy → Healthcare costs
- Program participation ← Education → Nutrition → Healthcare costs

There is still only one causal path; the path linking program participation to healthcare costs. The other paths are toward the treatment variable and are, therefore, noncausal paths. As a result, the association between treatment and outcome is likely the result of these three paths. The main causal path is only partially responsible for any association between the treatment and outcome.

In causal inference lingo, any association between a treatment variable and an outcome variable due to noncausal paths is called *confounding bias*. In other words, the noncausal paths create a bias in our estimates. The variables that are responsible for confounding bias are called *confounders*. In the figure above, being healthy, education, and nutrition are all confounders.

Sometimes, we are lucky, and there are no noncausal paths; therefore, there is no confounding. In most observational studies, however, we have confounding bias. We shall soon see that if we want the causal estimate of treatment on the outcome to be unbiased, we need to eliminate confounding bias by conditioning on the confounders. The first step in identifying confounding bias is to look at the causal graph that outlines our problem.

Blocking a Path

To understand how we can block a path, we must first understand a few concepts.

First, two variables are dependent if information about one gives us information about the other. In the following causal graph, education and income are dependent, as there is a (causal) link between the two.

Similarly, two variables are independent if information about one does not tell us anything about the other. For instance, while your level of education tells us some information about your income level, your shirt color does not tell us anything about your income. Note that if X1 is dependent on X2, then X2 is dependent on X1, and if X1 is independent of X2, X2 is independent of X1.

Second, blocking a path between two variables is equivalent to making those two variables independent of each other. By blocking a path, we intend to stop the flow of information from one node to another. Consider the following path.

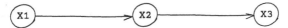

In this causal graph, causality flows from X1 to X2 to X3 and does not change direction. When the value of X1 changes, the value of X2 likely changes; when X2 changes, the value of X3 is expected to change. As a result, the value of X3 depends on the value of X1. But here is the catch. If we keep the value of $X2$ constant, then a change in $X1$ will no longer cause a change in $X3$. In other words, by keeping $X2$ constant, $X1$ and $X3$ become independent.

The concept above is what we mean by *conditioning on a variable*. By conditioning on a variable, we hold its value constant. In the causal graph above, if we want to block the path between $X1$ to $X3$, we hold $X2$ constant, also referred to as *conditioning on $X2$*. By doing so, $X1$ and $X3$ will no longer be dependent on each other. In other words, if we control for $X2$, having information about $X1$ tells us nothing about $X3$. We use the terms *conditioning on* something and *controlling for* something interchangeably.

The causal graph above is a special case. Information flows through a chain. Other forms of causal paths are forks and colliders. Let's consider each of these patterns in more detail. We'll start with chains.

Blocking a Chain Path

A chain is a path containing three nodes that takes you from the first node to the third node without changing direction. In the causal graph above, you can go from $X1$ to $X2$ to $X3$ without changing direction. As we saw, if we keep the value of the middle node, $X2$, constant, we block the path between $X1$ and $X3$.

Let's consider a conceptual example of a chain where $X1$ is inches of rain in an hour, $X2$ is the number of rides an Uber driver gets in an hour, and $X3$ is the income earned by the driver in that hour. If there is heavy rain one afternoon, we can expect the following sequence of events: more people will request rides, so the driver gets more rides, and the more rides the driver gets, the higher the driver's income will be. Here, $X1$ causes $X2$, and $X2$ causes $X3$. We can say that rain is not a direct cause of the driver's income but an indirect cause.

We said that if we hold X2 constant, X3 no longer depends on X1. If we have information about X2, knowing about X1 does not tell us anything about X3. What does this sentence mean in the context of our example? In plain English, if we know the number of rides requested, the driver's income is no longer determined by the amount of rain. To put it simply, a chain is blocked if we condition on the middle node.

What if the chain is longer? Consider the chain path shown next. How do we ensure that the path between X1 and X4 is blocked? Do we need to condition on every single middle node? The answer is no. By holding *any* of the middle nodes constant, we can make sure that the path is blocked. While we can condition on *all* middle nodes, conditioning on even one of them should do the job.

To see why conditioning on only X2 or X3 would make X1 and X4 independent, we can first focus on the chain X1 → X2 → X3. If we condition on X2, then X1 and X3 would be independent of each other. If X1 is independent of X3, then it will no longer be dependent on X4 either.

Blocking a Fork Path

Now consider a slightly different path, as shown below. In this graph, one node is the common cause of the other two. Here, X2 is the common cause of both X1 and X3. This is called a fork. It should be a no-brainer to see that if the common cause in a fork changes, the value of the other nodes will change. In this case, if the value of X2 changes, the values of X1 and X3 both change. So, by virtue of X2, X1 and X3 are dependent. This causal graph shows that X3 and X1 are not independent because the same information flows to both.

Can you guess how we make X1 and X3 independent? If the value of X2 does not change, X1 and X3 become independent. Conditioning on X2 makes X1 and X3 independent.

For example, think of X2 as the average income of a country, X1 as the annual chocolate consumption in said country, and X3 as the number of Nobel laureates from that country. We would naturally find a strong correlation (dependence) between annual chocolate consumption and the number of Nobel laureates in a country. However, chocolate consumption and Nobel laureates share a common cause: how rich a country is. The more affluent people are in a country, the more likely they are to consume chocolate. The richer people are in a country, the more likely the country is to produce Nobel laureates.

If we look at countries with the same income level, the dependency should no longer exist. In other words, if we condition on X2, the country's average income, X1, chocolate consumption, and X3, the number of Nobel laureates, will be independent. Like a chain, the fork pattern suggests that controlling the middle node should make the other two variables independent.

Blocking a Collider Path

A third and final type of path is a collider path. A collider is the opposite of a fork. In a collider, two end nodes are each a direct cause of the middle node. In the causal graph below, X1 and X3 are both causes of X2. Node X2 is called a collider, and the path is called a collider path.

A collider is different than chains and forks in that in colliders, the two end nodes (here X1 and X3) are automatically independent of each other, and there is no need for controlling for X2. In other words, because of how the nodes are connected, without holding X2 constant, information about X1 does not tell us anything about X3 and vice versa—because there is no flow of information from one to the other.

Here is when things get interesting about a collider. In the causal graph above, if we do condition on X2, X1 and X3 become dependent on each other. If we unintentionally condition on a collider—thinking we are removing confounding—we may introduce confounding in our analysis. A collider path is automatically blocked, and we do not need to condition on the middle node.

Why Are Collider Paths Automatically Blocked?

Consider the following example to understand how collider paths are automatically blocked without conditioning on the middle node. Suppose an educational grant is based on merit and need, that is, to qualify for the funding, the student has to be both in financial need and in good academic standing. The causal graph below shows the causal relationship:

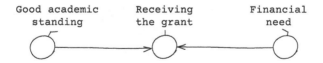

This causal graph is a clear example of a collider where financial need and academic standing affect receiving the grant, and the node receiving the grant is a collider. We can make this example very simple by assuming all the variables are binary:

- Good academic standing: 1 if the student is in good academic standing and 0 if otherwise,

- Financial need: 1 if in financial need and 0 if otherwise,
- Receiving the grant: 1 if the student receives the grant and 0 if otherwise.

To qualify for the grant—receiving the grant equal to 1—the value of both academic standing and financial need has to be 1. If either of the two nodes is 0, the student does not qualify for the grant, and, therefore, the value of receiving the grant will be 0.

If we do not condition on the collider node, receiving the grant, information about academic standing does not tell us anything about financial need or vice versa. One can be 0 while the other is 1; both can be 0, or both can be 1.

Now, let's assume, for whatever reason, we condition on the collider node, receiving the grant. By doing that, we hold receiving the grant constant and, therefore, we know its value. Let's say we know the student did not receive the grant—receiving the funding is equal to 0. Also, assume we know the value of financial need is 1. Then we know for sure the value of academic standing has to be 0; if both are equal to 1, the student automatically qualifies for the grant. By conditioning the collider, information about one of the root nodes gives us information about the other.

To wrap up, if we condition on the collider node, the path between the two root nodes is no longer blocked. We open that path that we should not have opened. If a collider is controlled for in a study by design or by mistake, the bias caused by creating an unwanted and noncausal path between treatment and outcome is called *collider bias*. In the next chapter, we discuss collider bias in more detail.

Identifying Confounding Bias in Observational Studies

As we saw before, one of the essential steps in conducting causal analyses or evaluating them is coming up with the causal graph that dictates the relationships between the important variables in the study. We then identify the noncausal paths and make sure they are all blocked. If there are unblocked noncausal paths, we identify the variables we need to control to block those paths.

But how do we come up with the causal graph in the first place? Although there are recent developments in empirically discovering causal graphs based on data—referred to as *causal discovery*—we usually use our expert knowledge and intuition to develop the causal graph. We do the detective work. Does variable X1 cause D? Does X2 cause Y? Or does Y cause X2 and X3? The answer to some of these is obvious. We may know that X2 cannot cause X1 just because, chronologically, X2 does not come before X1. For instance, parental genes affect a child's genes, not vice versa.

However, sometimes establishing what affects what is not as straightforward. You may argue that higher levels of antioxidants may lower the chances of heart disease, but other people may disagree with you. The best guide in discovering causal graphs is to use expert knowledge and previous research to claim any causal link. Suppose previous randomized experiments have found that giving higher doses of antioxidants reduced people's chances of heart disease. In that case, you can use that as evidence in constructing your causal graph.

A causal graph is likely a combination of many chains, forks, and colliders. We know that to block a chain or fork path, we need to condition on one of the middle nodes. A collider path is automatically blocked, so no action is required. Let's see blocking paths in practice.

In the causal graph on the following page, D is the treatment, and Y is the outcome. It is easy to see that there is a direct causal path from D to Y. There is also the noncausal path D ← X2 → X1 → Y. This noncausal path is a combination of a chain and a fork and does not include any colliders. Conditioning on either X1 or X2—or both—would block the noncausal path.

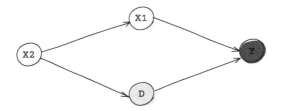

In the causal graph below, there is a direct causal path from D to Y. There is also the noncausal path D ← X1 → X2 ← X3 → Y. We need to block this noncausal path—but this noncausal path includes a collider, node X2, and this path is already blocked.[1] Hence, the association between D and Y only captures the causal path D → Y and not the noncausal path.

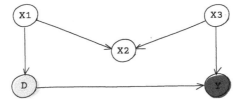

Let's make things a bit more complicated. The following causal graph includes a direct causal path and many noncausal paths. The noncausal paths include

- D ← X2 → X3 → Y
- D ← X2 → X1 → Y
- D ← X2 → X1 ← X3 → Y
- D ← X2 → X3 → X1 → Y

We aim to ensure that all of these noncausal paths are blocked and determine the minimum set of variables we need to control. Conditioning on X2 will block all paths as this node appears in all paths. The minimum set of variables that we need to condition on is X2, because conditioning only on X2 should block all paths. Be

---

[1]. Note that even if a path includes many more nodes, as long as one of the middle nodes is a collider, the path is blocked.

careful that in the third path, X1 is a collider, so if we condition on X1, we will create a path instead of blocking it. Therefore, another set of variables that would block all of the paths above would be X1 and X2.

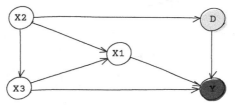

In the employer-sponsored wellness programs we discussed, let's assume that the following variables govern the causal dynamics. If participation in wellness programs is the treatment variable and healthcare costs in the outcome variable, then the noncausal paths include

- Program participation ← Being healthy → Healthcare costs
- Program participation ← Education → Nutrition → Healthcare costs

Based on the causal graph below, for the treatment effect of participation in wellness programs on healthcare costs to be unbiased, we need to condition on being healthy *and* either education or nutrition. We can also condition on all of these variables to make sure all noncausal paths are blocked.

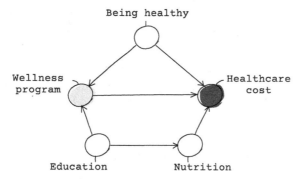

Using Regression Models to Reduce Confounding Bias

We have used the terms *conditioning on* or *controlling for* variables. We said that by conditioning on a variable, we keep its value constant. In the example above, where we need to condition on education and being healthy, we keep their values constant. We can find the association between participation in wellness programs and healthcare costs for fixed values of the variables being healthy and educated.

This method is called *stratification*. The idea of stratification is straightforward. Given that we have already identified confounders (being healthy and education), in stratification, we first stratify—read *categorize*—the data into strata based on these variables. Then, in each stratum, we can identify those who received the treatment and those who did not and calculate the treatment effect by comparing the average outcomes between the treatment and the control groups. By doing that, we will end up with multiple treatment effects, one for each stratum. The overall causal effect will be the weighted average of each treatment effect.

Unfortunately, stratification typically presents a set of problems. Stratification can lead to empty strata. If our sample size is small and the number of strata is large, there will be strata without any control or treatment observations, or both. In this case, our estimation based on stratification will be unstable.

Think about a study with one hundred subjects and only two main confounders: gender (with levels male or female) and age (with levels 0–9, 10–19, 20–29, etc.). If we stratify by these confounders, we might have strata—for example, male subjects ages seventy to seventy-nine—for which there are no observations. Even if there are observations in a stratum, there could be a case in which all of them are in the treatment group with no observation in the control group, or vice versa. We should, therefore, think about alternative methods. Some of those alternatives include regression, matching, or inverse probability weighting. We do not discuss these methods here.

Regression analysis requires further assumptions that we also do not discuss here. In general, regression is a better tool for dealing with continuous variables—as opposed to stratification—and it is better at extrapolating when we may have categories with no observations.

When we write a regression model as

Outcome = Intercept + c1 Treatment + c2 X1 + c3 X2 + Noise

we are finding the causal effect of treatment on the outcome by holding X1 and X2 constant. The treatment effect would be the coefficient on the variable treatment, or $c1$.[2]

Unobserved Confounding

In all of the examples above, we assumed that once we identify noncausal paths, if there are any, we make sure all the noncausal paths are blocked and the only remaining paths are the causal ones in which we are interested. By doing this, we find unbiased estimates of the treatment on the outcome since all spurious paths are blocked.

We also saw that blocking noncausal paths is done by conditioning on variables in paths that include chains and forks and doing nothing in paths that have colliders. But what if some of the variables we need to condition on are not available in the data? When a variable in a noncausal path needs to be blocked but is unobserved, we call that variable an *unobserved confounding variable*. If we must condition on that variable but our dataset does not include the variable, then we cannot eliminate the confounding bias caused by that variable.

---

2. This is because the partial differentiation of the outcome with respect to treatment is $c1$. In partial differentiation of a variable Y with respect to a variable D, we are holding all other variables constant, so here we are treating X1 and X2 as content.

72   CAUSE, EFFECT, AND EVERYTHING IN BETWEEN

In a situation like this, our causal estimate of the treatment effect is biased and needs to be taken carefully. There are methods for dealing with unobserved confounding. One of these methods is instrumental variables, where we find a variable as an instrument for unobserved confounding.

Another method for dealing with unobserved confounding is called sensitivity analysis. The premise of *sensitivity analysis* is to estimate the extent to which unobserved confounding is allowed to make our estimates statistically insignificant. To keep things simple, this book does not explore these more advanced methods. However, we are now familiar with the fundamental language for performing and critiquing causal analyses. The next chapter introduces a group of causal methodologies called quasi-experimental methods that help when estimating unbiased treatment effects is impossible due to the presence of unobserved confounding.

Chapter 5 Problems

1. Determine the path type and how to block it in the following scenarios. Draw a DAG if it helps. What does conditioning on one variable mean for the other two variables?
    a) When it is sunny, more people go outside. When more people go outside, restaurants make more revenue.
    b) Having more children affects family house size. Having more children also affects whether the family lives in urban or suburban areas.
    c) Countries with higher GDP per capita produce more pollution. Countries with higher populations produce more pollution.
    d) Increase in greenhouse emissions leads to increases in the global temperature. The change in temperature melts the glaciers, which in turn leads to rises in sea levels.
2. Consider the following DAG. The treatment variable is screen time, and the outcome is posture. List all the noncausal paths

and find a minimal set of variables to control for in order to block all noncausal paths.

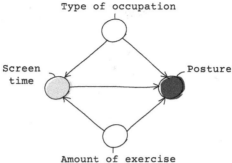

3. Consider the DAG from the practice question in chapter 3. The study is about the effect of playing video games on stress levels among youth. List all the noncausal paths and find a minimal set of variables to control for in order to block all noncausal paths.

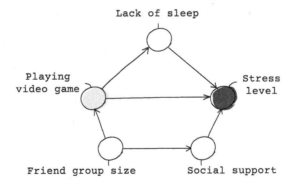

4. Consider the DAG from the practice question in chapter 3 about the effect of exercise on BMI. List all the noncausal paths and find a minimal set of variables to control for in order to block all noncausal paths.

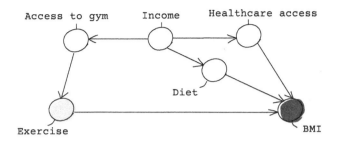

5. Consider the DAG from the practice question in chapter 3. The treatment variable is D and the outcome is Y. List all the noncausal paths and find a minimal set of variables to control for in order to block all noncausal paths.

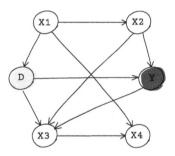

6. In the question above, how would the regression model that produces unbiased causal effects look like?

6

Quasi-Experimental Methods

Beyond experimental or observational studies, a third category of studies is receiving more attention in applied causal inference. These studies are called quasi-experimental studies.

Quasi-experimental methods help us use observational data in a way that mimics a randomized experiment. In these studies, the researcher does not have full control over the assignment of subjects to treatment and control groups. Unlike true randomized experiments, where random assignment is used to ensure that the groups are more or less similar, quasi-experiments rely on other methods to approximate this equivalence.

There are two ways we can have a quasi-experimental setting:

- A setting where even if the data are observational, the assignment to treatment is "almost" random. We call these quasi-experimental settings natural experiments, which we'll study in the upcoming section. In *natural experiments*, assignment to treatment and control groups is determined by external factors or natural occurrences rather than by the researcher.
- Another form of a quasi-experimental setting happens when the treatment to assignment isn't random. However, by using methods such as regression discontinuity, synthetic control,

or difference-in-differences, we can create situations where it is safe to assume that the treatment and control groups are similar or follow similar paths. In this chapter, we go over some of these methods.

Natural Experiments

In a natural experiment, the researchers have observational data that mimic a randomized experiment. While the researchers themselves have no control over which subjects fall into treatment and control groups, by chance, the subjects have been assigned to each group *as if* they were randomly assigned. Sometimes, the treatment assignment is caused by natural phenomena like weather patterns, date of birth, and genes, and sometimes by policies implemented through lotteries. Through this almost-random process of assignment, we can mitigate confounding bias. Let's dive into an example from the field of economics.

Does watching *Sesame Street* improve a child's future academic outcomes? A lot could go wrong in assessing this causal question. We can imagine access to the show being tied to factors that may also influence the outcome. One of these factors could be household income; it influences the likelihood of watching the show and also determines a child's future academic outcomes.

Fortunately, researchers Kearney and Levine were lucky enough to use data from a natural experiment.[1] *Sesame Street* mainly aired on stations affiliated with the Public Broadcasting System (PBS), which often broadcast on UHF (ultra-high frequency) channels. There are two reasons why families across the United States didn't have the same access to the show: First, many television sets at that time did not have the capability to receive a UHF signal. Second, UHF channels are, in general, weaker compared to VHF (very high

---

1. Kearney, M. S., and Levine, P. B. (2019). "Early Childhood Education yy Television: Lessons from *Sesame Street*." *American Economic Journal: Applied Economics, 11*(1), 318–350.

frequency) channels for physical reasons. This made the distance to the UHF antenna an important factor in determining who could watch the show. As a result, only about a third of American children could not watch the program even if they wanted to.

These technological constraints allowed Kearney and Levine to exploit the geographic variation in access to the show. If we think of those living in areas with a weaker signal as the control group and those living in areas with a stronger signal as the treatment group, we could argue that families were *naturally* randomized into treatment and control groups. They didn't have a choice over the strength of their signal for the most part. It is important to mention that a weak signal did not mean the children could not watch the show, and a strong signal did not mean they necessarily watched the show. The strength of the signal was just a proxy[2] for access to the show.

Using the natural setting above, the study found that children who could watch the show, especially those from disadvantaged families, benefited the most from the show; academically, they did better than those who received the weaker signal.

In some natural experiments, nature or the physical world does the randomization for us. For instance, Burke et al.[3] used randomness in rainfall around the world to study the effect of climate on civil conflicts. So, sometimes the setting in a natural experiment is indeed created by nature, and sometimes it's not. Be careful not to interpret the word "nature" or "natural" as only referencing nonhuman effects or interventions caused by nature. When we say "natural experiments," we are trying to highlight the fact that the treatment was assigned naturally without any involvement from the research team or the subjects themselves.

A perfect natural experiment is one in which the assignment to treatment and control groups is fully randomized. In other words,

---

2. A proxy variable in statistics is an indirect measure or substitute that approximates or represents a variable that is difficult to measure directly. It is used when direct measurement is not feasible due to constraints like cost, time, or unavailability of data.

3. Burke, M., Hsiang, S. M., and Miguel, E. (2015). "Climate and Conflict." *Annual Review of Economics*, 7(1), 577–617.

we can assume that those who ended up in the treatment group were just as likely to end up in the control group and vice versa. This natural assignment doesn't necessarily guarantee balanced treatment and control groups, but it is usually preferred to the case when subjects self-select into one or the other group.

In the following section, we go over a few more examples of natural experiments.

Long-term effects of in-utero influenza exposure during the 1918 Spanish Flu. At the end of the First World War, a flu pandemic killed more people around the world than the war. This flu was later known as the Spanish flu.[4] The earliest recorded case was in March 1918 in Kansas, with subsequent cases in France, Germany, and the United Kingdom by April. Over two years, nearly a third of the world's population, about five hundred million people, were infected across four waves. Death estimates typically range from seventeen million to fifty million, with some even estimating that deaths reached one hundred million, making it one of the deadliest pandemics in history.

In the United States, over the course of the pandemic, about one-third of women of childbearing age contracted the virus. Each wave of the pandemic was unpredictable and relatively short-lived, so prospective parents were not in a position to adjust their family planning in response to the flu. During the pandemic, some children were born to parents who were exposed to the virus, while other children were not. One could argue that these children were otherwise similar and that the chance of being born to an exposed mother was random. In other words, it is fairly safe to assume that the treatment and control groups were randomly assigned. No researcher imposed the virus on individuals involved in the study, making this a clear example of a natural experiment.

---

4. The term "Spanish flu" originated because the pandemic emerged near the end of World War I, a time when wartime censors in the countries involved in the conflict suppressed negative news to maintain public morale. In contrast, newspapers in neutral Spain reported freely on the outbreak, creating a false impression that Spain was the epicenter of the disease. This led to the misnomer "Spanish flu," despite the fact that the virus did not actually originate there.

Douglas Almond, an economist, used this setting to study[5] whether this in-utero exposure to the virus had any long-term health effects. The study found that individuals exposed to the flu in utero had lower educational achievements and lower income later in life. Additionally, poorer health outcomes and a higher incidence of physical disabilities were observed among those exposed to the flu before birth.

The Oregon Health Insurance Experiment. The Oregon Health Insurance Experiment is one of the most well-known natural experiments in public health and economics. The study was about Medicaid in the United States. Medicaid is a federal and state program designed to help with the healthcare expenses of low-income individuals.

In 2008, when the annual enrollment in Medicaid had closed nationwide, the state of Oregon realized it still had ten thousand spots available. So they decided to reopen enrollment for a limited time, not knowing that a lot more people would apply than the number of spots available. Roughly ninety thousand people applied for the remaining ten thousand spots. How did the state allocate the available spots among the applicants?

The state figured that a fair way of allocating the spots was running a lottery. In other words, they randomly allocated the spots to a subset of those who applied. These state officials had no intention of running a randomized experiment. Their intention was fairness in the allocation of Medicaid.

When the randomization was done, some researchers couldn't wait to have their hands on the data from one of the only large-scale natural experiments in healthcare. A group of researchers used the data to study the impact of access to Medicaid on health-related outcomes by comparing those who won the lottery to those who did not.

---

5. Almond, D. (2006). "Is the 1918 Influenza Pandemic Over? Long-Term Effects of In-Utero Influenza Exposure in the Post-1940 US Population." *Journal of Political Economy*, 114(4), 672–712.

Utilizing this near-perfect natural experiment, researchers were able to compare the outcomes of those who won the lottery to those who did not. What they found was that Medicaid significantly increased the use of healthcare services, including hospital admissions, emergency department visits, outpatient visits, and prescription drug use. The use of preventive care services such as mammograms also went up. Financial hardships related to medical expenses were diminished, with fewer unpaid medical bills sent to collection agencies and a reduction in medical debt, borrowing, and bill-skipping to cover medical costs. Catastrophic medical expenditures were nearly eliminated.

When it came to mental health, the depression rate in the group with access to Medicaid was 9 percent lower compared to those without Medicaid. However, and this is probably the most important finding of the study, access to Medicaid had no statistically significant effect on physical health measures such as blood pressure or cholesterol, despite an increase in diabetes diagnoses and medication use.[6]

Now that you know what natural experiments are, let's turn to the second category of quasi-experimental methods in causal inference. This second category consists of methods used to address the fact that the assignment to treatment and control groups has not been randomized.

In the following sections, we review three main methods that fall under this category: regression discontinuity design, synthetic control, and the difference-in-differences method. Although these forms of quasi-experimental studies are less compelling and harder to sell, they provide inventive ways of addressing confounding bias, especially in the presence of unobserved confounders.

---

6. You can read more about the experiment in: Finkelstein, A., Taubman, S., Wright, B., Bernstein, M., Gruber, J., Newhouse, J. P., Allen, H., Baicker, K., & Oregon Health Study Group. (2012, August). "The Oregon Health Insurance Experiment: Evidence from the First Year." *The Quarterly Journal of Economics, 127*(3), 1057–1106, https://doi.org/10.1093/qje/qjs020.

Regression Discontinuity Design

Nothing explains the concept of regression discontinuity better than an example. Let's look at one.

Imagine we're interested in the effect of college financial aid on college graduation. We want to know whether recipients of financial aid are more or less likely to graduate. To help answer this question we may consider conducting a randomized experiment. We ask a large enough group of college students to participate in our study; to some, we provide a financial aid package of $10,000, and to some, we don't. This will cost us a lot of money, and we also have to wait several years to compare the graduation rates between the two groups. What if we don't have the money or the patience to run a randomized experiment?

Let's say instead that we have observational data on an existing educational scholarship that awards financial scholarships to students based on their high school GPA. The scholarship is awarded based on a GPA cutoff point of 3.0. In other words, in order to qualify for the scholarship, a student's high school GPA should be above 3.0. This cutoff point is not chosen by us, the researchers or the students. It is instead determined by maybe a university or a government official. The point is that neither we nor the students chose the threshold.

The cutoff point cleanly divides subjects in our study into a treatment and a control group. Those above the cutoff point are the treated units that receive the scholarship, and those below are the control units that do not receive the scholarship. If you've been paying attention, you may immediately object to this setup on the grounds that it doesn't give us an apples-to-apples comparison between the treatment and control groups. The two groups could be different in many important ways. And you're right! Students with GPAs in the 2.0–3.0 range and students with GPAs in the 3.0–4.0 range might be fundamentally different from each other. The students in the former group might come from families with lower incomes. So, income might be a confounding variable.

But what if we look at students with GPAs just below and above the cutoff points—let's say those with GPAs in the range of 2.95 to 3.05? Can we say among the sample of students in this range that those who received the scholarship as a group are similar to those who didn't receive the scholarship?

It turns out that if we pick a small enough range around the cutoff point we can assume the two groups are similar.[7] Those above or below the cutoff point are arbitrarily in the treatment group. They didn't choose which side they were going to be on. So, if we only compare those just above to those just below the cutoff point, we can assume the two groups are very similar. As with a randomized experiment, when the two groups are similar, we can rule out the effect of any potential confounding variables.

This simple comparison of the two groups around the cutoff point is called a *regression discontinuity design*. Regression discontinuity resembles a localized randomized experiment because we can only study those around the cutoff point and not everyone in the sample. The word *regression* appears in regression discontinuity because researchers typically use a regression model to fit the data on either side of the cutoff point; *discontinuity* appears because the cutoff point causes a discontinuity in terms of treatment assignment. In a regression discontinuity study, the variable that determines whether an observation falls above or below a predefined cutoff point is called the *running variable*. In our example, our running variable is GPA.

Although regression discontinuity design is relatively simple in theory, we should carefully review the study design and interpret the results. In the following section, we review some of the main considerations when performing regression discontinuity.

---

7. This range is usually referred to as the *bandwidth*.

Some Considerations in Regression Discontinuity Designs

The continuity assumption. We discussed that a well-designed randomized experiment ensures that the treatment and control groups are, on average, similar in all aspects except for one: the treatment assignment. One group is randomly assigned to treatment, and the other group isn't. Similarly, because a regression discontinuity design is supposed to mimic a randomized experiment among subjects close to the cutoff point, we need to assume that the two groups are similar to each other. In other words, the continuity assumption is all about whether the characteristics of the subjects just above the cutoff point are, on average, similar to the characteristics of the subjects just below.

Going back to the example above, if we only include those in the GPA range of 2.95 to 3.05, we need to ensure that the characteristics of those in the GPA range of 2.95 to 3.0 are, on average, similar to the characteristics of those in the 3.0–3.05 range. For instance, if the average family income of one group is $55,000, we should expect the average family income of the other group to be close to this number. If 5 percent of students in one group come from single-parent families, the share of students with single-parent families in the other group should be close to 5%. This should be true for any variable we think is relevant to our study, and we can simply check the assumption if we have these *relevant* variables in our data.

This assumption can be violated in situations where the subjects are aware of the cutoff point and try to game it. In regression discontinuity, we assume that the cutoff point is not determined by the subjects of the study. An implication of this is that the subjects can't choose the side of the cutoff point where they end up. However, if, for some reason, subjects in a regression discontinuity design can choose which side of the cutoff point they end up on, the assignment to treatment within the bandwidth is no longer random, and we might have confounding bias.

To understand this, let's review one of the applications of regression discontinuity in evaluating a policy in Colombia. Camacho and Conover studied the effect of a welfare program in the early 1930s initiated by the Colombian government.[8] A *welfare program* is typically a government initiative designed to provide financial assistance and support to individuals and families in need. These programs usually have eligibility criteria. In this case, the government used a poverty index created for each household to determine eligibility for the program. Among other characteristics, the index was created based on the household's income, demographics, and the employment status of the head of the household. The index ranged from 0 to 100, and a household's index number would need to be 47 or smaller to qualify for the government benefit.

What Camacho and Conover discovered was that after 1998 the number of those who qualified for the program was larger than usual. What do you think happened?

While the formula for calculating the poverty index hadn't changed, some households manipulated their reported characteristics in a way that would make them eligible for the program. They found a way to game the system. In the context of regression discontinuity, the cutoff point was no longer exogenous, and people, to some extent, could choose their eligibility for the treatment. This behavior in regression discontinuity is usually called *sorting* or *bunching*.[9]

Is the cutoff used for another rule? Let's examine another potential problem. Suppose in the example above, the government uses the cutoff rule of 47 not only to determine eligibility for the welfare program but also for an education assistance program.

---

8. Camacho, A. and Conover, E. (2011). "Manipulation of Social Program Eligibility." *American Economic Journal: Economic Policy*, 3(2), 41–65.

9. This is a good example of Goodhart's law, which states that when a measurement, such as a poverty score, becomes a target, it stops being an effective measurement because people will alter their behavior to achieve the target.

Because in this case, the cutoff rule is used for multiple policies, looking at subjects around the cutoff points and comparing their average outcomes no longer tells us the treatment effect of the policy under the study. In this case, since the cutoff rule is used for determining eligibility for two programs, any effect we find could and should be attributed to both policies and not just the welfare program.

Choosing the right bandwidth. When it comes to regression discontinuity design, we usually can't use the entire sample in the analysis; we can only use those *just* above or below the cutoff point. In the student financial aid example, we couldn't compare those with a GPA of 2.0 to those with a GPA of 4.0. Our causal estimates would be valid only if we considered a small range around the cutoff point. But what range is considered small? Is looking at a sample of students with GPAs in the range of 2.5 to 3.5 good enough? Or should we make the range even smaller? How do we determine how wide or narrow the bandwidth should be?

Determining the bandwidth is not as straightforward as it sounds. While there are various methods for determining the *optimal* bandwidth, researchers usually show the treatment effect under various bandwidths and leave the judgment of which bandwidth is best for the reader.

Generalizability of the treatment effect in regression discontinuity. Something that is important in any causal study and even more important in regression discontinuity design is the generalizability of the treatment effects. The generalizability of a study is also called external validity. In causal inference studies, *external validity* refers to the extent to which the findings of a study can be generalized to other settings, populations, times, and circumstances beyond the specific conditions of the original study. In other words, external validity assesses the applicability and relevance of the study's results in real-world situations.

Remember that in regression discontinuity, we compare outcomes of subjects just below and just above the cutoff point. So even if the causal effect is correctly estimated, the results are only applicable to the observations within the bandwidth. Sadly, it's not uncommon to find examples where study results are overextended and misused to infer conclusions about the broader population.

Generally, you should frame your results with modesty. Don't generalize! Don't exaggerate! Be specific about the population to which your causal estimates apply. This is especially important in regression discontinuity design. In the student aid example, we found the effect of financial aid on the graduation rate to be 0.05 percentage points when we look at students with a GPA bandwidth of 2.95 to 3.05. In this case, we can't extend the treatment effect to subjects outside of this GPA range.

Synthetic Control Method

Before we delve into the synthetic control method, we need to know about panel data. *Panel data*, also known as *longitudinal data*, refers to a dataset that observes the same subjects over multiple time periods. Panel data allow researchers to not only take advantage of cross-subject variations in the data, but also variations over time. An observational panel dataset is a panel dataset on a set of subjects without manipulating the environment or the subjects being studied. The following table shows a simple panel dataset tracking population and GDP per capital over five-year intervals for various countries.

The synthetic control and difference-in-differences methods are methods of causal inference that work with panel data. Let's start with the synthetic control method. The idea behind synthetic control is simple.

In synthetic control, there is usually only one treated observation. The synthetic control method helps us weigh observations in

TABLE 6.1 An example of a panel dataset

Country	Time	Population (Millions)	GDP per capita
Angola	2000	13.5	$801
Angola	2005	16.9	$2,257
Angola	2010	19.6	$4,540
Cuba	2000	11.2	$2,307
Cuba	2005	11.3	$4,327
Cuba	2010	11.2	$5,539
...	...	...	...
Zimbabwe	2000	11.8	$586
Zimbabwe	2005	12.9	$424
Zimbabwe	2010	13.0	$753

the control group (which could be many), so that together they form an equivalent observation for the treated observation. The control group observations then form one single subject comparable to the treated observation. In other words, this single synthetic subject is the counterfactual for the treated subject. By simply comparing the synthetic control unit to the treated unit, we can then find the treatment effect.

Maybe an example would illustrate this better. In the early 1990s, Sweden implemented a carbon tax on transport fuels, becoming one of the first countries to do so. The tax started at $30 per ton of CO_2 and has since risen to $132 per ton, making it the highest carbon tax in the world.

Researcher Julius Andersson sought to understand the causal effect of Sweden's carbon tax on carbon emissions.[10] If the effects were significant, such a policy could be recommended to other countries. However, Andersson faced the challenge of dealing with

---

10. Andersson, J. J. (2019). "Carbon Taxes and CO2 Emissions: Sweden as a Case Study." *American Economic Journal: Economic Policy,* 11(4), 1–30.

observational data. A simple before-and-after comparison wouldn't be convincing due to numerous other factors influencing the environment during those periods. Similarly, comparing Sweden to other countries wouldn't be effective due to potential confounders. Isolating the causal effect of the tax in this case proved to be difficult.

We could use the synthetic control method. We could collect data on ten European countries that did not implement a carbon tax in the 1990s. Let's say that Norway, Denmark, France, and the United Kingdom are some of these countries. We can use these observations to build a single synthetic control subject; let's call it synthetic Sweden. Our synthetic Sweden is like a stew; it has a little bit of Norway, a little of Denmark, a pinch of France, some United Kingdom, and so on and so forth.

The purpose of the synthetic control is to represent what would have happened to Sweden had it not implemented the policy. Therefore, by comparing Sweden to "synthetic Sweden," we can estimate the causal effect by subtracting the outcomes from each other.

The set of control subjects that constitute the synthetic control unit is usually called the *donor pool*. The starting point for a synthetic control analysis is to find a reasonable donor pool—reasonable because it's better that the donor pool subjects are as similar as they can be to the treated unit. For instance, if we use in the donor pool a set of countries that are economically, politically, and geographically different from Sweden, our synthetic control won't be very useful. The main challenges associated with the synthetic control method is to identify appropriate units to include in the donor pool and determine weights for each donor.

The data we use for the synthetic control method should include some relevant variables for the donor pool and the treated units. Julius Andersson used GDP per capita, the number of motor vehicles, gasoline consumption per capita, and the percentage of urban population as key variables. The algorithm for finding the weights has one objective: The pretreatment outcome trend of the synthetic control unit should be similar to the pretreatment outcome trend of the treatment unit. Here pretreatment refers to the time period before the treatment happened (the period before Sweden introduced its carbon tax).

Andersson used twenty-four countries as donor subjects and found the weights for each of the twenty-four countries in the donor pool as follows. Unsurprisingly, Denmark had the highest weight at 0.48, meaning it contributed 48 percent to synthetic Sweden and was the closest match in terms of pretreatment confounders. New Zealand followed with a weight of 0.17, then the United Kingdom at 0.13, and Belgium at 0.1. The remaining countries in the donor pool had really small weights.

Once we find weights that make the pretreatment trends similar, we can use these weights to calculate the posttreatment outcome trend for the synthetic unit. In the end, any differences in the posttreatment outcomes between the treated and the synthetic control unit can be attributed to the treatment effect.

The following graph, taken from Andersson, shows that the pretreatment trends in metric tons per capita of CO_2 between Sweden and synthetic Sweden are very similar. Therefore, the treatment effect at each point in time is the difference between the posttreatment outcomes of the synthetic and treatment units in each year. For example, in the year 2000, the treatment effect is roughly 0.5 metric tons per capita. The following shows the trends for per capita CO_2 emissions from transport between 1960 and 2005 for Sweden and synthetic Sweden.

As observed, following the introduction of the policy, Sweden's pollution rate diverged from its synthetic counterpart. Posttreatment emissions in actual Sweden were lower than those in synthetic Sweden, indicating that the policy effectively reduced emissions.

The synthetic control method has been employed to investigate various topics, including the impact of cigarette taxes in California and the economic cost of unification in West Germany. It is

90 CAUSE, EFFECT, AND EVERYTHING IN BETWEEN

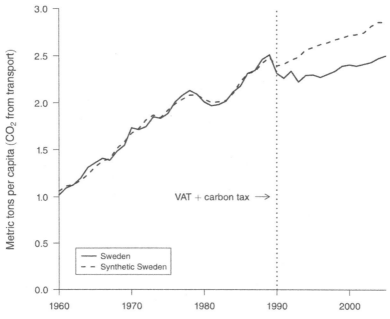

FIGURE 6.1 Metric tons per capita of CO_2 for Sweden and synthetic Sweden. Andersson, J. J. (2019). "Carbon Taxes and CO_2 Emissions: Sweden as a Case Study," *American Economic Journal: Economic Policy*, *11*(4), 1–30. Copyright American Economic Association; reproduced with permission of the *American Economic Journal: Economic Policy*.

primarily used in comparative studies involving small samples, a single treated unit, and multiple time periods.

However, the method has some limitations, which are as follows:

1. The selection of the initial donor pool is susceptible to various subjective biases. Researchers sometimes start with a large donor pool without adequately justifying why each control unit was included, which is poor practice. Each unit in the donor pool should share some similarities with the treated unit.

2. Additionally, the units in the donor pool must not have received treatment during the study period. In his carbon tax paper, Andersson excludes countries that implemented carbon taxes in the transport sector (Finland, Norway, and the Netherlands) and those that significantly altered fuel taxes (Germany, Italy, and the United Kingdom) during the sample period. Units that experienced large idiosyncratic shocks in their outcome variables during the study period should also be excluded from the donor pool.
3. In relation to the donor pool units and the treated unit, there shouldn't be the possibility of spillover effects, that is, the treatment on the treated unit should not affect the control units. As a hypothetical example, if the tax policy caused Swedish drivers to fill their tanks in Denmark, this assumption would be violated.
4. Lastly, the data need to be from multiple time periods. In fact, the weights of the control subjects are calculated based on their pretreatment trends, which need to be long enough for good weight calculation. The synthetic control method is a before-and-after treatment analysis that requires data from both before and after the treatment for both the treated and control subjects.

Difference-in-Differences

We saw that the synthetic control method requires panel data for estimating causal effects. However, there are generally two limitations. One is that the method can handle only one treated unit. We can imagine many scenarios where the treatment group includes many more units. Another limitation is that the method requires a long enough pretreatment period to calculate properly the weights for the donor subjects. So, when we have a dataset that only has one or a few pretreatment time periods, we need to look for other methods.

Another method for estimating causal effects using panel data is the method of difference-in-differences. Let's go through an example using minimum wage data to understand this method.

The minimum wage is the lowest compensation that employers can pay their workers. Many countries have minimum wage laws, and in some countries, the minimum wage varies from geography to geography. Modern minimum wage laws were first introduced in New Zealand and Australia in the 1890s.[11]

While the minimum wage aims to ensure a basic living standard for low-wage workers, economic supply and demand models suggest that increasing it might lead to job losses. Empirical evidence, however, remains inconclusive on whether raising the minimum wage actually results in higher unemployment rates.

Answering this question causally is difficult because randomized experiments are impractical in this context. It is challenging to persuade local and federal governments to conduct such experiments, leaving us reliant solely on observational data.

Imagine you have data on the United States, broken down by state, including a variable indicating whether a state implemented a minimum wage increase (MWI) and another variable representing the employment rate in each state.

We know that a simple comparison of employment rates between states that pass an MWI and those that don't will not work. States that pass an MWI are different from the states that don't in unobserved ways; for instance, the political climate across the states may be quite different.

Card and Krueger, two renowned labor economists, saw this as a prime opportunity to apply the difference-in-differences method. Their paper is now regarded as a classic example of this approach in action.[12] So what exactly did they do?

---

11. Starr, G. (1993). *Minimum Wage Fixing: An International Review of Practices and Problems.* 2nd impression, with corrections. Geneva: International Labour Office, 1.

12. Card, D., and Krueger, A. B. (1994). "Minimum Wages and Employment: A Case Study of the Fast-Food Industry in New Jersey and Pennsylvania." *The American Economic Review*, 84(4), 772–793.

Their study focuses on New Jersey, where the statewide minimum wage was raised from $4.25 an hour to $5.05 an hour in 1992. Although this increase may seem modest in absolute terms, it represented an 18 percent increase. Meanwhile, in neighboring Pennsylvania, there was no change in the minimum wage. Card and Krueger took advantage of this difference by collecting data from individual businesses in the fast-food industry, where many minimum wage workers are employed. They compared the outcomes in New Jersey to those in Pennsylvania before and after the minimum wage increase in New Jersey, using data from two points in time: one before and one after the increase.

The employment rates in both New Jersey and Pennsylvania are naturally different from each other. Let's consider the scenario that the trends move in a parallel manner, i.e., if employment rate in one goes up (down), the other goes up (down) by the same amount. The idea of difference-in-differences is that under this scenario, we could assume that in the absence of the policy (treatment), the trends would remain parallel. However, because of the treatment, any deviation from the parallel trend in the treated group can be attributed to the treatment effect.

The difference-in-differences method tells us that we can measure the treatment effect by capturing how much deviation we have from the parallel trends. The method is called *difference-in-differences* because the treatment effect can be calculated by first looking at the differences between the two trends pretreatment and then subtracting them from any posttreatment deviation from the parallel trends.

The strength of difference-in-differences is that by taking the pre- and posttreatment differences, we eliminate individual effects (selection bias), and by calculating the treatment-control difference, we eliminate time effects. This effectively minimizes the influence of unobserved confounders.

Unlike the synthetic control method, which aims to make the synthetic control unit as similar as possible to the treated unit, the difference-in-differences method only requires the trends in the treated and control units to move in a parallel fashion. This is the method's key assumption, and it is a strong one. This assumption is called the *parallel-trends assumption*.

Let's estimate the causal effect in our example. Card and Krueger found the following employment rates:

TABLE 6.2 Minimum wage data for Pennsylvania and New Jersey

Time	Pennsylvania	New Jersey	New Jersey – Pennsylvania
Before MWI	23.33	20.44	–2.89
After MWI	21.17	21.03	–0.14
After – Before			2.75

The numbers in the table above display full-time employment in both states before and after the MWI. Estimating the treatment effect from this data should be straightforward.

1. Before MWI, the difference in employment rates between New Jersey and Pennsylvania was –2.89. In other words, the employment rate in Pennsylvania was 2.89 percentage points higher.
2. After MWI, the difference in employment rates between New Jersey and Pennsylvania went down to –0.14.
3. Because we assume that in the absence of treatment, the employment rate difference between the two states should have remained around –2.89, any deviation from this number can be called the treatment effect.
4. So we can simply take a difference between the two differences and call it the treatment effect. The difference-in-differences causal estimate is then the difference between the two, which is $0.59 - (-2.16) = 2.75$.

This result indicates that the estimated causal effect of the minimum wage hike in New Jersey was a 2.75-percentage-point increase in the employment rate. In other words, the minimum wage increase actually boosted employment rather than hindering it. The publication of this paper marked a significant shift in many economists' views on minimum wage laws. It highlights the impact of causal inference–based research, as the paper empirically challenges prevailing economic theory.

It is noteworthy to mention some of the important criticisms of the paper. Critics had two major complaints:

1. **Selection of before-and-after points.** The minimum wage increase in New Jersey took effect in 1992, but it was announced two years earlier. This gave employers in New Jersey roughly two years to adjust their employment practices. If employers did change their behavior in advance, the comparison of employment just before and after 1992 may not be as insightful as intended.
2. **Parallel-trends assumption.** New Jersey and Pennsylvania both experienced a small recession during the study period. Even if this recession impacted employers in New Jersey and Pennsylvania similarly, the responses of employees in each region might have differed. There is evidence that while employers in Pennsylvania let workers go, employers in New Jersey raised their menu prices (the businesses examined by Card and Krueger were fast-food restaurants).

Wrapping Up Quasi-Experimental Methods in Causal Inference

In this chapter, we explored various quasi-experimental methods: natural experiments, regression discontinuity design, the synthetic control method, and difference-in-differences. While there is so much more to learn about these methods, my hope is that these brief introductions give you a solid footing to explore each method further.

The central theme across these methods is that answering causal questions typically requires a randomized setting to address confounding bias. However, in the absence of enforced randomization, researchers seek quasi-experimental settings, where

1. The treatment assignment is naturally randomized (like in natural experiments).
2. The treatment assignment is not random, but either an equivalent control group for the treated group can be found or constructed (as in regression discontinuity and synthetic control) or we can assume that the control and the treated units move in a parallel fashion in the absence of the policy (as in difference-in-differences).

These methods should not be confused with fully randomized experiments. Sometimes a quasi-experimental study can be convincing; other times, it may seem less credible, and you might feel that the researchers are reaching to advance their arguments. To be able to fully understand these methods, perform them, or be a better critic of them, you need to delve deeper into each method.

When evaluating a quasi-experimental study or conducting one yourself, it is essential to prioritize the underlying assumptions over the statistical calculations. Pay close attention to the eligibility criteria: who qualified for the treatment and who did not. Consider whether subjects self-select into treatment and control groups in ways that are not observed. Identify any variables that might be relevant for selection into the treatment group and could serve as potential confounders.

7

A Framework for Evaluating Causal Studies

We are going to organize everything we have learned so far into a framework for evaluating causal studies you see around you. Open any scientific magazine, newspaper, or social media links, and you will see many causal claims around you. The guidelines you learn in this chapter hopefully equip you to go beyond the headline. This framework helps you evaluate the elements of a causal analysis.

Is the Study Causal?

The first step is to make sure we are dealing with a causal claim. One way of figuring out if a study is causal is to look for causal language in the study text. Is the study making causal claims, or is it only suggesting an associative—correlational—relationship? In chapter 2 we learned about some of the identifier words in correlational and causal studies. The headlines below are all from popular media and respectable newspapers. What is common among them?

- Air pollution linked to far higher COVID-19 death rates, study finds.
- Protein tangles in Alzheimer's patients could help predict brain shrinkage.

- Israel fish deaths linked to rapid warming of seas.
- Study links cannabis use during pregnancy to autism risk.

The common theme is that they mainly point at associations. Saying that two variables are linked together, correlated, or associated together is not equivalent to saying that one causes another. Similarly, if we say that A predicts B, we mean A and B are associated. For instance, saying that being tall predicts higher income does not mean that it causes you to have higher income. It merely points to the fact that information about one tells you something about the other. Associative claims are more humble and easier to prove.

So what is the importance of associative or correlational studies? Establishing a correlation between cannabis use during pregnancy and autism in children may not necessarily mean that cannabis use among mothers causes autism. It may simply raise an alarm that further research is needed. Maybe mothers with autism are more likely to use cannabis but are also expected to pass genes responsible for autism to children. Perhaps cannabis use and alcohol consumption are linked, and it is, in fact, alcohol consumption that causes autism. Maybe mothers who use cannabis during pregnancy have mental health risk factors that are also responsible for autism in their children. We see that correlational studies do not answer casual questions. Instead, they raise casual questions.

Importantly, associative language is not prescriptive because it does not tell us what we should do. If a study suggests that cannabis and autism are correlated, the finding should not lead to an actionable prescription. Because prescriptive studies are more actionable, researchers sometimes get tempted to use causal language without having done a causal analysis. A causal claim is a lot more attention-grabbing than an associative claim. Look at the following headlines and decide what is common among them:

- E-cigarettes are effective at helping smokers quit, a study says.
- Is conference room air making you dumber?

- Having more sex makes early menopause less likely, research finds.
- The sex of researchers affects the language of research papers.
- Zapping the brain improves creativity.
- Diets rich in tea, berries, and apples could lower blood pressure.
- Going vegetarian may lower risk of UTIs in women, study finds.
- Video gaming can benefit mental health, find Oxford academics.
- High blood pressure and diabetes impair brain function, study suggests.
- Hyperventilating can help clear alcohol from the body faster, researchers find.
- Racism ages Black Americans faster, according to our 25-year study

The words used in these studies have more of a prescriptive nature and contain causal language; pay attention to the words *help, make, affect, improve, lower, benefit,* and *impair*. These statements are prescribing something to the reader based on the causal claim. Saying that zapping the brain improves creativity is prescribing something to the reader. Saying that video gaming can benefit mental health, the researcher claims that video games do something. They cause better mental health. Whether or not the causal studies behind the headlines above are sound, they are making causal claims.

Sometimes the causal language is hidden and not as explicit as other times. Look at the headlines below:

- Children raised in greener areas have higher IQ, study finds.
- Infants exposed to air pollution have less lung power as adolescents.
- Genes play a role in the likelihood of divorce.
- Women who were tall and lean in childhood are more at risk of endometriosis.

These headlines avoid explicit causal language but nonetheless imply causality. Saying that genes play a role in divorce suggests that genes are a causal determinant of divorce. To most people, saying that children raised in greener areas have higher IQs means greener air may cause higher IQs. So a first step in evaluating a study is determining whether it implies any causality or is merely a correlational study. If they are making causal claims, then proceed to dig deeper.

What Are the Treatment and Outcome Variables?

If a study is causal, it should have an explicit treatment variable. Sometimes the treatment variable can be manipulated, like vitamin intake or job training. We can increase or decrease these treatment variables. Sometimes the treatment variable cannot be manipulated even if the study is causal. For instance, consider the headline about the effect of height and leanness on endometriosis. The treatment variables are being tall and lean. While leanness can be manipulated, we cannot manipulate height.

As we discussed, sometimes the treatment variable is continuous, and sometimes binary or discrete. Getting versus not getting a vaccine means dealing with a binary treatment variable. However, we can have a study where the treatment variable is different vaccine dosages. It can be 2.5 mg, 10 mg, or any number in between.

The same goes for the outcome variable; we should always clearly state the outcome variable in a study. I should note that some studies look at multiple outcomes. For instance, a causal investigation of immigrants' education on their assimilation in the host society might look at numerous factors as proxies for assimilation. It may look at financial connectedness, marriage, and fluency in the host language.

Like the treatment variable, the outcome variable can be binary or continuous. For instance, a study on the effect of a vaccine on getting a disease has a binary outcome variable. Individuals in our study either get the disease or do not.

Finally, the unit of measurement of the treatment and outcome variables should be clearly stated. One of the studies above investigates the effect of video gaming on mental health. We should ask ourselves how video gaming is measured. Is it playing as opposed to not playing video games? Or is it measured in terms of hours of playing video games? Or minutes?

Furthermore, how is mental health measured? In terms of having or not having a diagnosed mental health illness, or is it on a scale of 1 to 10? Is it self-reported or physician-reported? These are all critical questions to ask. The interpretation of the causal effect depends on the scale of these variables.

Is the Study Sample Representative of the Population?

A joke in statistics goes like this: we conducted a phone survey and found that 98 percent of respondents love responding to phone surveys. Inference about a population depends on the sample on which the inference is based. In evaluating causal studies, we need to investigate whether the sample used in the study is representative of the population to which the causal claims apply. An important rule is that the sample selection should be convincing. An excellent way to guarantee that a sample is representative of a population is when it is drawn randomly from the population. For instance, if we are sampling the voting population in the United States, it is not okay to only contact subjects through landlines because younger people are less likely to use landlines. It is also not okay to only contact subjects on X (previously Twitter) because older people are less likely to use it. A good sampling strategy would ensure that everybody in the voting population has a chance to appear in our sample.

Additionally, we need to look further for the external validity of the results. External validity refers to whether study results are likely to hold over variations in study settings, population, treatment, and outcome. Even if a study was done well, will the results hold if we tweak the treatment or choose a different demographic or locality?

Here is an example. In 1984, two criminologists ran a randomized experiment to investigate what type of police response to calls about domestic violence led to better outcomes. In collaboration with the Minneapolis police, officers were randomly assigned one of the three treatments: arrest the assailant, advise the couple, or send the assailant away for eight hours. The study concluded that the arrest option led to lower repeat violence reported.[1]

The study received much attention since it laid out a framework for local police to deal with domestic violence anywhere in the United States or even as a prescription for other countries. However, some researchers were suspicious that the results might not necessarily apply to different settings. They asked if there was something specific to the demographic or the community investigated in the study. The only way to answer this question was to replicate the study in other localities. Unfortunately, the results did not hold under other replications.

The researchers who investigated the study found that mandatory arrest led to lower repeated violence in communities considered more stable. In such communities, the public shaming of the arrest would make the assailant less likely to repeat the crime. However, in less stable communities, public arrest did not produce the same results.[2]

So it is vital to identify the population to which the result of a causal study applies. Hasty generalizations can mislead readers into thinking that causal effects apply to all settings.

---

1. Sherman, L., and Berk, R. A. (1984). *The Minneapolis Domestic Violence Experiment.* Washington, DC: National Policing Institute. https://www.policinginstitute.org/publication/the-minneapolis-domestic-violence-experiment/

2. For a summary of the critique, read Binder, A., and Meeker, J. W. (1993). "Implications of the Failure to Replicate the Minneapolis Experimental Findings." *American Sociological Review, 58*(6), 886–888.

If an Experimental Study, Is It Done Well?

An essential distinction in evaluating a study is whether it is experimental or observational. If the study is experimental, then our focus should mainly be on whether the randomized experiment was done well or not. However, if the study is observational, we usually focus on whether there is confounding that would bias the results.

There is usually a clear statement in a causal study about whether it was a randomized experiment or not. Different disciplines may use different terminology for randomized experiments, such as randomized controlled trials (RCTs) or clinical trials.

If the study is experimental, it should use randomization for assigning subjects to treatment or control. In most cases, we do not have to worry about if randomization is good enough, but as we saw before, one of the ways that randomization can become problematic is when there is noncompliance. When noncompliance exists, the researcher either needs to control for confounding factors related to noncompliance or use an approach for calculating treatment effect by using the share of compliers in the study.[3]

As discussed, another issue that can happen in randomized experiments is attrition. Similar to noncompliance, attrition, if non-negligible and nonrandom, can lead to confounding bias. As a result, you should look for a significant attrition rate and, if one is present, whether the study researchers tried to address it.

Lastly, an issue that can arise in randomized experiments and observational studies is the spillover issue we have discussed. In causal studies, we need a clear distinction between subjects in the treatment and control groups. A subject's treatment should not affect the outcome of any other subject in the study. Let's take a look at an example.

Imagine a study examining the effect of using face masks on contracting the SARS-CoV-2 virus. The study probably suffers from

---

3. We call a treatment effect estimated in the presence of noncompliance *local average treatment effect*.

interference or spillover issues because when the treated subjects in our study wear masks, it likely affects the virus's spread, affecting the outcomes for everybody. As a result, it becomes hard to distinguish who received treatment and who did not.

In general, we should worry about interference in contexts where subjects can interact with each other. A study may suffer from interference in the presence of general-equilibrium effects. To understand general-equilibrium effects, think of distributing basic income as the treatment to some city residents. In the long term, distributing basic income means the average income in the city is higher, which then affects aggregate unemployment and inflation for everybody in that city.

If an Observational Study, Is It Done Well?

If the study we are evaluating is observational, our main concern should be confounding bias. Unfortunately, few studies provide a causal graph for your review to identify the noncausal paths. Therefore, to check for confounding bias in an observational study, think of the causal graph that makes sense to you. You should then identify the noncausal paths, as we saw in the previous chapter, and come up with the minimum set of variables that need to be controlled.

Once you know what the study should control to block all noncausal paths, check whether the researchers have indeed controlled for those variables. If your causal graph is correct, and the study comes short of the confounding variables list, there is potential confounding bias.

Statistical Significance versus Practical Significance

While we did not go over statistical significance in this book, a brief mention does not hurt. Suppose the treatment effect we find in an experimental or observational study is greater than the treatment effect by chance. In that case, we call the treatment effect statistically significant. If 95% of the time, the treatment effect is larger than that obtained by chance, we say the treatment effect is statistically significant at 95 percent. If a treatment effect is only 6 percent or 30 percent, or even worse, 80 percent of the time larger than what we find by chance, then we question our treatment effect, and we call the treatment effect statistically insignificant. In these cases, the estimate that we find may not fully be due to the treatment and could also be due to chance.

Sometimes, however, we may find a statistically significant treatment effect, but the magnitude is not substantial. For instance, evaluating a tutoring program for college students, we may find a statistically significant average treatment effect of 0.03 units increase in students' GPA. Even if the effect is statistically significant—larger than what we find by chance most of the time—according to the program evaluator or the study researcher, the 0.03 increase in GPA may not be large enough to justify an investment in the tutoring program. So, besides wanting statistically significant treatment effects, we want treatment effects that are large enough to be justified.

What is large enough? The answer to this question depends on the context and expectations. In some contexts, an 0.03 increase in GPA might be justifiable, especially if the program is not too expensive. In some cases, it is not. In some government programs, an increase of $500 in families' wealth may be large enough, and in some cases, it may not be.

A better way to understand whether a treatment effect is sizeable is to look at the number in relative terms. For instance, imagine that a treatment effect is a five-unit decrease in the outcome variable. Is this large or small? Regardless of the context, it would help if you asked what the range of the outcome variable is. If the outcome variable varies largely, say from zero to 1,000, then a 5 unit change is small. However, if the outcome variable varies within a narrow range, like between zero to ten, then a 5 unit change would translate to a 50% increase which is large.

Is There Any Potential Collider Bias?

In the previous chapter, we saw that a noncausal path that contains a collider is already blocked and does not impose confounding bias. However, if a study conditions on a collider by design or mistake, a noncausal path is created, leading to bias. We called this bias the collider bias. There are two ways we can have collider bias.

Collider bias by mistake. In observational causal studies, researchers commonly consider any variable relevant to the outcome and treatment, and condition on them in the analysis to be safe. This practice is widespread when the researchers ignore constructing a causal graph before the study. Among these variables that are related to the outcome and the treatment, there might be a collider. So by mistakenly conditioning on a collider, researchers can accidentally induce collider bias. This situation is, in fact, not rare.

Collider bias by design. Collider bias can also happen by design. This collider bias is a more complicated case and is sometimes referred to as sample selection bias. Let's review an example that helps us understand this collider bias.

Do GPAs determine success at big companies? This question is based on the claim made by Laszlo Bock, senior vice president of people operations at Google, who said in an interview that looking at Google employees, they found that GPAs are "*worthless as a criterion for hiring.*"[4] In other words, Google employees with low GPAs are as good as Google employees with high GPAs. We can use our causal tools to see what is behind this claim and whether it makes sense.

The causal question concerns the effect of GPA on workplace efficiency. Let's go through the causal dynamics. First, we can't ignore the fact that GPA may influence the likelihood of being hired by a company like Google. Someone during the hiring process may look

---

4. See https://www.nytimes.com/2013/06/20/business/in-head-hunting-big-data-may-not-be-such-a-big-deal.html

at the GPA line on an applicant's resume. So we can claim that GPA affects being hired by Google. At the same time, productive people have a higher chance of being hired by Google, so job performance affects being at Google. Let's look at this simple causal graph.

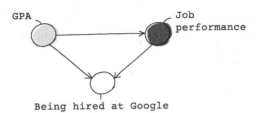

Being hired at Google

The noncausal path in this graph is GPA → Being hired by Google ← Job performance, and this path contains a collider since the two end nodes collide in the middle node. So far, so good. If being hired by Google is a collider, the noncausal path is already blocked, and there is nothing to sweat over. The claim by the Google vice president is acceptable.

Here is the catch, though. This claim is based on Google employee data, that is, everyone in this study is at Google. In other words, we are keeping the variable working at Google constant, which means we are controlling for it. Our sample is already conditioned on the collider, which means we have a spurious noncausal path between treatment and outcome open. There is collider bias, and Bock's claim should be taken with a grain of salt. This type of collider bias is sometimes called *sample selection bias*; we have a bias in our estimates because of how we sample data. In this case, instead of making a claim about a general population, the claim is about a selected sample of only Google employees.

Checking for potential collider bias in a study is, therefore, very important. While collider bias is more common in observational studies, it can also happen in experimental studies. In a randomized experiment, the chance of causing collider bias by mistake is low since there is usually no need to condition on any variable. However, it is likely to have collider bias due to sample selection bias, as we discussed. In the next chapter, we review some causal studies that suffer from biases, including confounding and collider bias.

Now that we have some protocols for evaluating causal studies, it is time to put our knowledge into practice and review some actual studies. We try to cover a wide range of disciplines and scientific fields so that you can have a more general understanding of evaluating causal studies.

8

Causal Case Studies

This book started by defining causal inference, considering randomized experiments versus observational studies, investigating the different types of biases that plague causal analyses, and in the previous chapter, laying out a framework for evaluating causal studies.

In this chapter, we use this framework to assess causal claims from the Web, media, and academic papers. In none of these studies is my aim to critique the discipline nor the theoretical foundations, but rather the causal method and the causal conclusions.

Does Growing a Beard Have Anything to Do with Sleep Apnea?

Sleep apnea is a sleeping disorder in which breathing stops and restarts many times during one's sleep due to the closing of the airways in the throat. If untreated, apnea can lead to serious health problems such as high blood pressure and heart disease. It is estimated that about 15 to 30 percent of males and 10 to 15 percent of females in North America suffer from sleep apnea, making apnea a common disorder.

In the United Kingdom, sleep apnea affects around 1.5 million people, but it is claimed that 85 percent of the cases are undiagnosed. To help people self-diagnose if they have apnea, the British Snoring and Sleep Apnoea Association (BSSAA) has provided a series of questions, such as whether you breathe through your nose or mouth and are tired or have dry mouth when you wake up in the morning. Believe it or not, one of the questions asks, "Do you have a beard?"[1]

What does having a beard have anything to do with apnea? One hypothesis is about CPAP machines. A Continuous Positive Airway Pressure (CPAP) machine is a method of treating apnea. While you sleep, a CPAP machine keeps the airways open through a continuous pressurized stream of air delivered via tubes and masks. A beard may be a sign of untreated apnea because the gap between the mask and face caused by having a beard may let out some pressurized air, which leads to lower efficacy. But the test by BSSAA is for those not diagnosed with apnea yet, so most of them may not even know what a CPAP machine is.

So the answer should be something else. The other possible hypothesis is that some men grow a beard to cover their double chin, usually a sign of being overweight. We also know that being overweight will likely cause or worsen sleep apnea. Let's put these in a nice little causal graph.

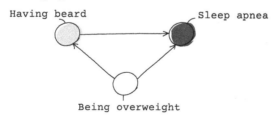

Based on what we know so far and the causal graph above, it seems like even though having a beard may have some direct effect on sleep apnea—something that I seriously doubt—it has an indirect effect through being overweight. Based on this, a better

---

1. https://britishsnoring.co.uk/itests/

question that BSSAA can ask is a direct question about BMI or being overweight.

Vaccination and COVID Mortality

In 2021, the *New York Times* published a plot that showed the relationship between the share of the total population fully vaccinated in each state in the United States and the death rate in those states. The scatter plot showed a negative correlation between the two, indicating that states with lower vaccination rates had higher death rates and vice versa.[2] The plot was widely circulated on social media. Paul Graham, a computer scientist and a startup guru, reshared the plot with the caption "Wow, a small difference in vaccination rates makes a big difference in fatality rates. The Covid fatality rate in Wyoming is about 4x that of Connecticut."[3] While the original article by the *New York Times* did not make any causal claims, the tweet suggested that vaccination, the treatment variable, is highly effective in reducing mortality rates from COVID, the outcome variable. Stating that one variable "makes a big difference" in another is typically interpreted as implying a causal relationship. Regardless of our prior knowledge and beliefs about vaccines, let's evaluate the causal claim in the tweet.

The dataset is observational as nobody randomized states into different vaccination rates in the hope of measuring the outcomes. Also, note that the units of analysis in this study are states and not individuals. Since we are dealing with an observational study, let's focus on confounding bias. Is there any? To answer this, we start with a causal graph and throw in any variable we think might be relevant.

---

2. Lu, D., and Sun, A. (2021). "Why Covid Death Rates Are Rising for Some Groups," *The New York Times*, December 28, https://www.nytimes.com/interactive/2021/12/28/us/covid-deaths.html.

3. Graham, P. (2021). X (Previously Twitter), https://x.com/paulg/status/1476556224594427907.

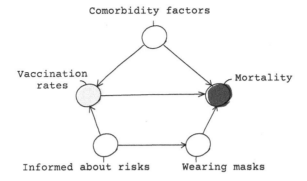

This causal graph is a simple one featuring only a few relevant variables. It assumes that comorbidity factors like obesity and diabetes affect mortality. Also, people with comorbidity factors may be more or less hesitant to get the vaccine—being more at risk affects getting the vaccine. The graph also emphasizes the effect of information about COVID. It assumes that being informed about the risks associated with COVID makes people more likely to get the vaccine but also more likely to wear masks and take other cautionary measures. Wearing masks and being cautious can, in turn, lower the risk of dying from COVID. We can proxy information about risks of COVID by the level of education in each state.

If these assumptions are valid, we must block two noncausal paths:

- Vaccination rates ← Share of people with comorbidity factors → Mortality
- Vaccination rates ← Share of educated people → Share of people wearing masks → Mortality

How do we block these paths? By conditioning on the middle nodes. To block the first path, we need to condition on comorbidity factors, and to block the second path, we need to condition on either the education level or share wearing masks or both. The causal claim and the accompanying graph in the post do not suggest considering any of that; therefore, we are dealing with confounding bias, and the causal conclusion mentioned in the tweet is problematic.

Having Work Best Friends and Job Satisfaction

A 2006 Gallup book[4] outlined twelve elements of great management, and the tenth element –encouraging friendships in the workplace– drew particular attention. A *Washington Post* columnist wrote, "A best friend at work? . . . What is this? High school?"

The book uses Gallup's survey data to support its 12 elements. Based on the survey done in 150 countries, 30 percent of respondents reported having a best friend at work. The book then goes on to say that these people were seven times more likely to be engaged at their jobs.

Saying that those who are X are more likely to be Y is a purely correlational statement. For example, observing that cities with more ethnically diverse restaurants tend to have a higher share of educated residents does not mean that ethnic food causes higher education levels—or vice versa. Rather, an internationally diverse food scene may serve as a useful predictor of a city's education level.

The statement in the Gallup book was deliberately cautious, avoiding claims that workplace friendships cause greater engagement and productivity. However, later posts by Gallup and other media stretched beyond this careful correlational framing. Another Gallup blog post[5] headlined "Why We Need Best Friends at Work" quoted the same findings. A *New York Times* opinion piece from 2022[6] used the results from the Gallup survey to hint at the statement that "friendships benefit productivity." A blog post on a company's website said, "Why you should have a best friend at work." All of these statements interpret the

---

4. Gallup (2006). 12: The Elements of Great Managing, https://store.gallup.com/p/en-us/10242/12:-the-elements-of-great-managing.

5. Mann A. (2018). Why We Need Best Friends at Work, Gallup, https://www.gallup.com/workplace/236213/why-need-best-friends-work.aspx.

6. Goldberg, E. (2022). The Magic of Your First Work Friends. International New York Times, https://www.nytimes.com/2022/07/14/business/work-friends.html.

correlational findings as causal. If an increase in X benefits Y, X is causing Y.

Aside from what the original findings suggested and the subsequent interpretations, having friends at work may, in fact, benefit engagement at work and more productivity. Having friends at work may make people happier, more satisfied, engaged, and more productive. But there could also be other irrelevant channels at work. For instance, it could be that more sociable people tend to have more friends at work. Maybe they get jobs because their friends recommended them to that company. Additionally, more social people may be less stressed and, as a result, more satisfied and more productive at work. The following causal graph summarizes these arguments.

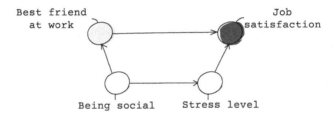

You should be able to think of other factors at play, but this simple causal graph is enough to show that aside from the direct causal path, there are noncausal paths that need to be blocked. A simple correlation between having best friends at work and job satisfaction also includes this noncausal path.

Does More Screen Time Lead to Behavioral Changes among Kids?

We have seen a similar example of the effect of screen time on obesity. Beyond obesity, there are also claims that spending too much time on screens has changed behavioral changes such as change in our attention span. Behavioral changes are said to be exceptionally prominent among teens.[7] The title of a UNICEF parenting resource page says, "Babies Need Humans, Not Screens," and the World Health Organization and the American Academy of Pediatrics caution against screen time among children.[8] Notably, experts caution against how screen time impacts mental health among children. Are these claims correct?

Let's first think about what screen time means as the treatment variable. If you were to pick a measure for screen time, which of the following would you choose? Screen time as a binary variable, meaning parents allow or do not allow screen time, or screen time as a continuous variable, meaning hours spent in front of screens? The second metric is more informative, as the number of hours is an essential piece of information.

Furthermore, we want to know a little more about the nature of screen time; one hour of spending time on the screen watching *Sesame Street* is not the same as one hour playing video games as regards the effect on mental health. It is, therefore, natural that different studies find different treatment effects.

The outcome variable is also potentially different across studies. Some may use the prevalence of anxiety, while others use aggressive behaviors or depression as the outcome.

Additionally, these studies are always observational; we cannot force people—let alone kids—into treatment and control groups regarding screen time. As a result, some studies may suffer from confounding bias. What could be a confounder here? Assume the outcome of interest is depression and the treatment variable is hours spent on screen. Each unit is a child five to ten years of age, and the sample is representative of the five- to ten-year-old-child population in the United States. Here are some of the assumptions or claims concerning the causal dynamics of these studies:

---

7. Hu, J. C. (2020). What Does a Screen Do?, Slate, https://slate.com/technology/2020/03/screen-time-research-correlation-causation.html.

8. UNICEF (n.d.). Babies Need Humans, Not Screens, https://www.unicef.org/parenting/child-development/babies-screen-time.

- The SES of the family, in terms of education and income, is a determinant of depression, but so also are the amount of screen time, hours of exercise, and hours spent socializing. The latter claim is based on the assumption that children from wealthier parents have more opportunities and time for socializing with family and friends.
- Hours spent socializing is a determinant of depression.
- Exercising affects depression.
- Screen time affects hours spent socializing and hours of exercise.

You can think of many other variables and connections, but this is a reasonable set of variables. The following causal graph is built based on the claims and assumptions just presented.

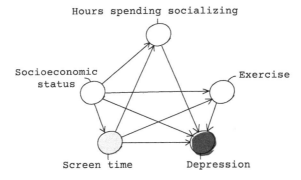

Assume that we are only interested in the direct effect of screen time on depression. Therefore, any indirect causal or noncausal path is undesirable and should be blocked. These would be the paths:

- Screen time ← SES → Depression
- Screen time ← SES → Exercise → Depression
- Screen time ← SES → Hours spent socializing → Depression
- Screen time → Hours spent socializing → Depression

- Screen time → Hours spent socializing ← SES → Depression
- Screen time → Hours spent socializing ← SES → Exercise → Depression
- Screen time → Exercise → Depression
- Screen time → Exercise ← SES → Depression
- Screen time → Exercise ← SES → Hours spent socializing → Depression

At first glance, it seems like we need to control for many variables to block all of the paths above. But there is a trick. One node is common to all the paths above. That node is SES and it does not seem to be a collider on any of the paths. So conditioning for SES, and SES alone, should block all the paths. In other words, if the causal graph that we came up with is reasonable, then any study on the causal effect of screen time on depression should adjust for the family's socioeconomic status to avoid confounding bias. Note that we may find other sets of confounders, but SES is the minimum set of confounders.

Does the Race of the Victim Matter in the Probability of Receiving Capital Punishment?

Michael Radelet, in his 1981 paper *"Racial Characteristics and the Imposition of the Death Penalty,"*[9] shows that those accused of murdering whites are more likely to be sentenced to death. However, when we condition on the offender's race, the disparity in receiving capital punishment between blacks and whites vanishes. These findings are from data between 1976 and 1977 from 600 homicide indictments in the United States, a few years after the 1972 Supreme

---

9. Radelet, M. L. (1981). Racial Characteristics and the Imposition of the Death Penalty. *American Sociological Review*, 918–927.

Court decision in *Furman v. Georgia* that supposedly eliminated racial disparities in capital cases. So what is the truth? Are homicides with white victims more likely to lead to capital punishment for the offender?

The treatment variable here is the victim's race, and the outcome variable is receiving the death penalty. The causal path that we are interested in is the path between the treatment and the outcome. Additionally, the race of the offender affects the race of the victim. Whites are more likely to kill whites, and blacks are more likely to kill blacks. The race of the offender affects the outcome variable due to systemic racism. The simple causal graph would look like the one below when ignoring other possible variables.

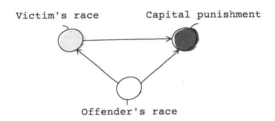

This causal graph shows a single noncausal path: victim's race ← defender's race → capital punishment. If we want an unbiased estimate of the effect of treatment on the outcome, we need to block this path. Hence, the results after conditioning on the defender's race are more trustworthy.

Does Having More Children Make Couples Happier?

A 2021 study in the academic journal PLOS One had the headline "When and How Does the Number of Children Affect Marital Satisfaction? An International Survey." Let's dig deeper and evaluate the study from a causal inference perspective.

The paper's title makes it clear that we are dealing with a causal study. However, in the abstract, the authors change gears and tone down the causal language by saying, "We found that the number of children was a significant negative predictor of marital satisfaction."[10] Factor X is a predictor of factor Y does not imply that X causes Y.

The treatment variable in this study is the number of children, and the outcome variable is self-reported satisfaction. The study looked at more than seven thousand married individuals from thirty-three countries and was an observational study. The sample is large and seems to be representative of the populations in those thirty-three countries. Researchers seem to have controlled for important variables such as sex, education, and religiosity. However, there is a shortcoming: the study only looked at individuals who were married at the time of the survey. Let's look at the causal graph.

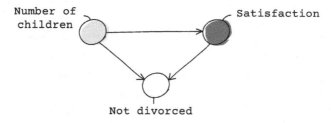

Why is looking only at married individuals in this study problematic? Since the study only looks at married individuals, it is, in effect, conditioning on being married—or not being divorced. We know that marital satisfaction determines getting divorced. Dissatisfaction with marriage leads to higher rates of divorce. But also, the number of children within a family may increase the chances of divorce, maybe due to financial strains caused by having more children. If that is the case, being married or not getting divorced is a collider, and by conditioning on this collider due to the nature of the sample, we create a noncausal path that creates bias. This example is

---

10. Kowal, M., Groyecka-Bernard, A., Kochan-Wójcik, M., & Sorokowski, P. (2021). When and How Does the Number of Children Affect Marital Satisfaction? An International Survey. *PloS one, 16*(4), e0249516.

another case of collider bias or, more specifically, sample selection bias.

Does Drinking Alcohol Increase the Chances of Heart Disease?

When it comes to alcohol and ischemic heart disease, there is not much of a consensus among researchers. Ischemic heart disease, also called coronary heart disease (CHD) or coronary artery disease, refers to heart problems caused by narrowed heart arteries that supply blood to the heart muscle. Wallach et al. show that different studies, primarily observational, have controlled for various confounding variables, leading to a wide range of treatment effects.[11] They show that no two papers controlled for the same set of confounding variables.

So which study is right? What is the *right* set of confounding variables? As it turns out, none of these studies employed a causal graph to figure out the right set of confounding variables, so there is a gap to be filled. Wallach et al. found that most studies used the following variables in covariates: smoking, which is the variable that appeared the most; age, body mass index or BMI; physical activity; and education. Let's see how these variables fit in a causal graph.

The treatment of interest is alcohol-drinking status, and the outcome of interest is having ischemic heart problems. As we saw, we need to make certain assumptions about how variables are related to come up with a causal graph. These assumptions should be based on prior research or common sense. If research tells us that physical activity affects ischemic heart disease, we will add an arrow

---

11. Wallach, J. D., Serghiou, S., Chu, L., Egilman, A. C., Vasiliou, V., Ross, J. S., and Ioannidis, J. (2020). "Evaluation of Confounding in Epidemiologic Studies Assessing Alcohol Consumption on the Risk of Ischemic Heart Disease." *BMC Medical Research Methodology*, 20(1), 1–10.

from physical activity to the outcome node. These are some of the assumptions we are going to make.

- The level of education affects how much a person drinks.
- The level of education affects how much a person smokes.
- The level of education affects how much a person exercises.
- Age is a determinant of alcohol consumption.
- Age is a determinant of smoking.
- Age is a determinant of having heart disease.
- Age is a determinant of physical activity.
- BMI is a factor in developing heart disease.
- Physical activity affects having heart disease.
- Physical activity affects BMI.
- Alcohol consumption affects smoking. Alternatively, you can assume it is the other way, meaning smoking affects alcohol consumption, but this did not change the final set of confounders.

You may agree or disagree with these claims but explicitly mentioning these assumptions is a good start. If these assumptions are reasonable, the resulting causal graph will look like the one below.

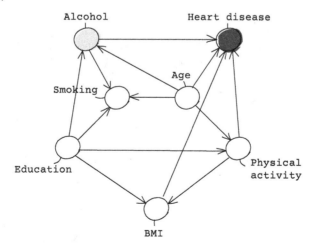

This causal graph is probably the most complicated one we have seen in this book. As it turns out, the graph has a lot of noncausal paths. There is a high chance that we will miss a few if we want to come up with all the noncausal paths. Luckily, software can help us find the noncausal paths that need to be blocked. They even return the minimum set of variables to prevent confounding bias. One of these tools is DAGitty.[12] You can quickly draw the causal graph and automatically have the results. After drawing the DAG, you can check the section called "Causal effect identification" on top of the right side panel for the minimal sufficient set of confounders as shown in the image below.

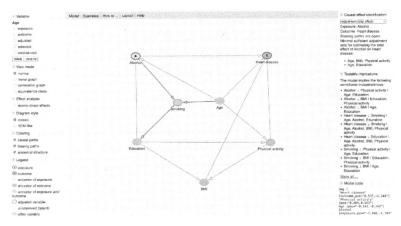

FIGURE 8.1 A screenshot of DAGitty web application

Using DAGitty, we find that conditioning on either set of variables shown below should block all noncausal paths.

- BMI, age, physical activity
- Age, education

12. Dagitty (n.d.). DAGitty — Draw and Analyze Causal Diagrams, https://www.dagitty.net/.

Conditioning on all the variables may not be the best strategy, as it may cause collider bias. However, we can trust studies that control for one of the above sets.

A Final Note

You may have learned many aspects of causal inference and possibly forgotten some along the way. However, you should now have enough knowledge to challenge your less causally savvy friends and family when they make or cite dubious causal claims.

I intentionally left out many other tools and variations of the methods from this book to avoid overwhelming you during your first journey in causal inference. Hopefully, you feel encouraged to continue your journey independently. If so, feel free to send a note and let me know about your progress. You should be able to easily find me on the Web.

ANSWERS TO END-OF-CHAPTER QUESTIONS

Answers to Chapter 1 Problems

1. Causal inference involves rigorous empirical investigations into causal questions and claims. In causal inference, we base our conclusions on expertise, causal reasoning, and empirical evidence, which we use to identify and estimate the magnitude of causal relationships. In our day-to-day lives, we tend not to be very rigorous in the way we answer causal questions, and we are often persuaded by causal claims that are convenient, sound good, are beneficial, or come from a place of authority. We are also more susceptible to fallacies like the post hoc fallacy and questionable-cause fallacy.
2. In causal inference, D is said to be a cause of Y if Y responds to changes in D. D does not have to be the sole cause of Y to be considered a cause.
3. In a deterministic definition, a cause D will always exert a change in its outcome Y. In other words, if D changes, Y must change. In a probabilistic definition, D is a cause of Y if it causes Y *most of the time*. We use the probabilistic definition in causal inference.
4. No, not necessarily! We can determine the exact value of Y if D is the only cause of Y. However, Y can have multiple causes. To know the exact value of Y, we need to know how each of the causes affects Y.

5. Although the question is not directly stated as a causal question, it implies that studying for an example affects exam results. By asking this question, we want to know the question "What is the effect of the time spent studying on the exam results?" So if we rephrase it that way, it will be a causal question.
6. An association (or correlation) shows that knowing something about one variable gives us some knowledge about another variable, but it doesn't imply any causal relationship between the two variables. Causation, on top of establishing a relationship between two variables, implies that one variable causes the other one to change.
7. No! Based on these facts alone, we cannot conclude that the lower interest rates caused the economy to grow. By jumping to this conclusion, we would be guilty of the post hoc fallacy. The facts tell us there is a correlation between interest rates and economic growth, but we would need to investigate this association further in order to establish a causal link.
8. Confounding bias is caused by outside factors that influence both the cause and effect.
    a) Student motivation. More academically driven students are more likely to use flashcards and more likely to have better learning outcomes.
    b) Weightlifting. Individuals who weight lift may be more likely to follow an all-meat diet and have higher muscle mass.
    c) Socioeconomic background. Those who come from a higher socioeconomic background are more likely to have more education and are also more likely to have a higher income level.
    d) Parents' education and income level. More educated parents and parents with higher income levels may be more likely to read to their children each night. Parents' education and income also directly impact a child's IQ as environmental factors can contribute to IQ.

e) Stress levels. Those who have lower levels of stress may be more likely to take yoga classes and also more likely to have fewer mental health issues.

Answers to Chapter 2 Problems

1. The following indicates which statement is considered causal and which is considered correlational.
    a) Causation: The word *improve* is an action verb. We can replace it with the verb *affect*. The sentence states that music lessons cause better brain development.
    b) Causation: The statement claims that housework *decreases* or *lowers* breast cancer risks.
    c) Association: Framing this way usually implies a simple causation. This sentence states that lucky people are born in summer but makes no causal claims.
    d) Association: The term *linked to* is usually interpreted as *associated with*. A *link* between two variables doesn't have any direction and doesn't imply any causation. The statement simply means that trust in government and work attitudes are linked.
    e) Causation: The sentence suggests that later school times cause a reduction in car crashes. Typically, the terms *reduces, decreases, increases, improves, affects,* and *influences* imply causation.
    f) Association: This statement suggests that there is an association, but no causation is established yet. The reason that this statement is a correlational statement is that we can replace the verb *identifies* with the verb *predicts*, and therefore, the statement falls under predictive problems that are in nature correlational.
    g) Causation: Although the term *cost* isn't among any of the key identifiers of causal statements that we have discussed, the statement implies that checking phones in lectures causes lower grades. Here the uncertainty implied by the word *can* refers to the magnitude of the effect.

h) Association: This one may be obvious to most people. The sentence doesn't tell us if men cause dogs to be more aggressive or that men typically adopt dogs that are already aggressive. So the statement is simply pointing to an association between the two. Here, no causation is claimed.

i) Causation: The sentence can be rephrased as "sincere smiling causes longevity," and therefore would likely be interpreted as a causal statement by most people.

j) Association: Deep-voiced men are associated with more kids. Again, the statement does not claim any causality.

2. Statements (a), (b), (e), (g), and (i) are the causal statements. For these statements, the treatment and outcome variables are as follows:

   a) Statement (a): The treatment is *music lessons* and the outcome is *brain development*.
   b) Statement (b): The treatment is *housework* and the outcome is *breast cancer risk*.
   c) Statement (e): The treatment is *school start time* and the outcome is *car crashes*.
   d) Statement (g): The treatment is *checking phones during lectures* and the outcome is *grades*.
   e) Statement (i): The treatment is *sincere smiling* and the outcome is *longevity*.

3. Some of the treatments in these statements can't be assigned, and some of the studies are considered unethical.

   a) We can theoretically assign people not to wear helmets while cycling, but that would be unethical.
   b) It is not possible to assign gender.
   c) It is theoretically possible to assign provision of milk at schools. It may be argued that running a study in which some schools are deprived of free milk would be unethical.
   d) We can assign daily school recess time or length, and this is more or less ethically possible.

e) We cannot change parents' socioeconomic status during childhood if the individual is already at college age, so this is not possible.
f) We can assign cell phone usage, and here there is no ethical issue.

4. For (a), the subjects would be cyclists (individuals). For (b), the subjects would be athletes (again, individuals). For (c), schools are potentially the subjects. In (d), children (individuals) are the subjects. Questions (e) and (f) both suggest individuals as the subjects of the study.

5. Only the first and the third statements can be interpreted as causal. The following
   a) This statement claims causation based on the use of the word "improve." If we want to change this statement to claim an association, we could say that the mental well-being of older adults is linked to social support.
   b) The word "linked" used here means that the authors claim an association. We could transform this statement into a causal one like this: drug use leads to child anxiety. Note that the claim could also be in the other direction: child anxiety causes drug use.
   c) Categorizing this one as causal or noncausal is a bit tricky. Generally, the use of action verbs such as "cuts" implies causation, but the word "seems" may add uncertainty about the causal claim and may change the causal interpretation. A clear causal statement would be that fans in rooms cut infants' risk of crib death. A clear noncausal statement would be that fans in rooms and infants' risks of crib death are associated with each other.
   d) This is a noncausal statement. One possible causal version could be that tooth loss causes mental impairment. Because correlational statements don't have any direction, we could also say that mental impairment causes tooth loss.
   e) This is a noncausal statement. A causal version of the statement would be that being an obese girl decreases the chances of going to college.

6. Again, only the first and third statements can be interpreted as causal. In the first statement, the treatment variable is social support. It can be measured using an index that combines several self-reported dimensions such as the number of friends, hours per week spent socializing, number of times going out with friends, and so on. These variables can be discrete or continuous. The outcome is mental well-being. Researchers can use evaluations and tests. The variable can also be self-reported. It will most likely be a discrete variable. In the third statement, the treatment variable is likely whether there is a fan in the room, so in that sense, it is a binary (discrete) variable. The outcome, infant mortality, is measured as a binary (discrete) variable indicating whether the child died in a crib.

Answers to Chapter 3 Problems

1. The DAG has a loop: D → Y → X3 → D. It is not acyclic and, therefore, not a valid DAG.
2. The following DAGs illustrate the causal studies:
   a) Scenario (a)

   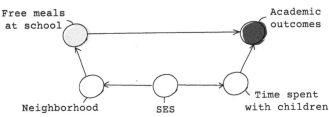

   b) Scenario (b)

   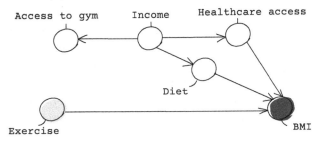

3. The causal graph suggests that the hours of playing video games affect stress levels both directly and indirectly. This may be due to the fact that playing video games in itself can cause stress but it can also cause stress due to changes in sleeping habits. Based on the DAG, the size of one's friend group affects hours of playing video games but also affects the level of social support youth receive. Finally, having a better social support system influences stress level.

4. All the possible paths are:

- D → Y
- D → X3 → X4 ← X1 → X2 → Y
- D → X3 → Y
- D → X3 ← X2 → Y
- D ← X1 → X2 → Y
- D ← X1 → X2 → X3 ← Y
- D ← X1 → X4 ← X3 ← Y
- D ← X1 → X4 ← X3 ← X2 → Y

The first four paths are outward of the treatment variable, so they are causal.

Answers to Chapter 4 Problems

1. The more samples we draw, the closer both groups should be to having one-third of mothers who have a college degree. Due to the small sample size, in the study with only twenty participants, we may not get a perfect balance, but with the larger study we would expect roughly one-third of mothers in each control and treatment group to have a college degree. This is under the assumption that the two dice are fair. Note that in an observational study, due to self-selection into treatment and control groups, we wouldn't necessarily get the same share of college-educated mothers.

2. We can imagine a ten-sided die or a random integer generator between 1 and 10. If the number is between 1 and 4, then the mother's eye color is brown; if the number is between 5 and 7, then it's blue; if the number is 8 or 9, then it's brown,

and otherwise, it's blue. We again throw a second die that captures the assignment to treatment or control. If the number is between 1 and 3, the mother is assigned to the treatment, and if the number is between 4 and 6, the mother is assigned to the control. If we do this experiment with a large sample size—that is, if we throw these two dice one hundred or five hundred times, we would expect the distribution of eye colors in the treatment group to be similar to the distribution of eye colors in the control group and similar to the distribution in the population. In other words, the variable eye color should be balanced across the treatment and control groups.

3. Answer (c) is the most helpful answer. Answer (b) is wrong since the increase is an average and does not have to apply to every individual in the sample, let alone to those outside of the study sample like your friend. Answer (a) does not provide enough information. I would tell her to consider her personal situation since the treatment effect might be different for different subgroups such as gender, race, education, or age groups.

4. The DAG would look like the one below. Note that since treatment is randomized, nothing affects the treatment variable. More specifically, access to a gym doesn't affect how many hours per week a person exercises. As a result, the only path that is between the treatment and the outcome would be the main causal path, and there won't be any noncausal paths between the treatment and the outcome.

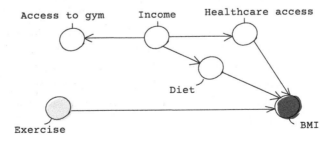

5. Because of the randomization, no other variables besides the coin toss influence the treatment variable. Therefore, the only path between the treatment and the outcome is the direct causal path.
6. The following are some of the shortcomings of randomized experiments in the context of the studies above:
    a) For this study, we cannot randomize race. Race is not something we can assign to people in a study.
    b) Randomization would be unethical since we can't randomize people to not wear seat belts and put their lives in danger.
    c) In this study, we are interested in the long-term effect of female elementary school teachers since the outcome is measured in adulthood. This would technically be what is called a *longitudinal study*, where the same subjects are followed over time. In these studies, attrition could be an issue.
    d) Here, noncompliance could be problematic. A lot of people drink coffee regularly and might have problems complying if they are assigned not to drink coffee.
    e) In this study, informed consent could affect the results of the study. Most people want to appear kind, and if they know that their kindness is being assessed, they might try to be more kind than they usually would be.

Answers to Chapter 5 Problems

1. The following illustrates the DAGs for each scenario.
    a) As we can see from the DAG below, this scenario is a chain. Because the path is a chain, sunny weather is correlated with restaurant revenue. We can block the path by controlling the number of people going outside. After blocking the path, sunny weather and restaurant revenue are now independent and not correlated anymore.

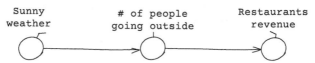

b) This scenario indicates a fork. Because of this, house size and living in suburban versus urban areas are dependent. We can change that by controlling for family size (number of children). This blocks the path and makes house size and living in suburban versus urban areas independent of each other. Note that this would happen if there were no other paths between house size and living in urban areas.

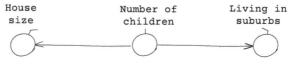

c) The DAG shows that we are dealing with a collider path. The path is blocked without controlling for the middle node, pollution; GDP per capita and population are independent. We have to be careful because controlling pollution unblocks the path and makes GDP per capita and population dependent on each other.

d) Even though more variables are involved this time, this is a chain. This means that all the variables are dependent on each other, for example, greenhouse emissions and rises in sea levels. Controlling for any of the variables in the middle of the path will block the path and make them independent. For example, we could control for changes in temperature to make greenhouse emissions and the rising sea levels independent of each other.

2. There are two noncasual paths in this DAG:
    - *Screen time $\leftarrow$ Amount of exercise $\rightarrow$ Posture*
    - *Screen time $\leftarrow$ Type of occupation $\rightarrow$ Posture*

   Both paths are forks. To block both paths, we have to control for amount of exercise and type of occupation. Therefore, the minimum set of variables we need to condition on is two.

3. Here, the only noncausal path is hours of playing video game $\leftarrow$ friend group size $\rightarrow$ social support $\rightarrow$ stress levels. To block this path, controlling for either friend group size or social support would be sufficient. So the minimum set of confounders is either:
    - {Friend group size}, or
    - {Social support}

   It should be noted that since social support is hard to measure, it may be a better strategy to control for the size of one's friends group. Additionally, as we saw, there are two causal paths here. One is direct and potentially the one we are interested in friend, and the other is indirect and through hours of sleep. If we are only interested in how playing video games directly affects stress levels, we should block the other causal path. In this case, we should control for the hours of sleep as well.

4. There are only two noncausal paths here. The two paths are:
    - Hours exercise $\leftarrow$ access to gym $\leftarrow$ income $\rightarrow$ access to health care $\rightarrow$ BMI
    - Hours exercise $\leftarrow$ access to gym $\leftarrow$ income $\rightarrow$ healthy diet $\rightarrow$ BMI

   Since access to gym and income are on both of these paths, controlling for either one of these variables would block all backdoor paths. So the minimum set of variables in this case would be either access to gym or income.

5. As we saw, there are four noncausal paths in this DAG.
    - $D \leftarrow X1 \rightarrow X2 \rightarrow Y$

- D ← X1 → X2 → X3 ← Y
- D ← X1 → X4 ← X3 ← Y
- D ← X1 → X4 ← X3 ← X2 → Y

The last two paths include the collider X4, so they are already blocked. The third path includes the collider B, so it's already blocked. To unblock the first path, we can either control for X1 or X2. So the minimum set of variables would be either X1 or X2.

6. As we saw, the minimum set of confounders is either X1 or X2. So, a regression model that includes either one of these variables would block all backdoor paths.

    Y = Intercept + b1 × D + b2 × X1

    Or,

    Y = Intercept + b1 × D + b2 × X2

    Equivalently, we could also control for both variables:

    Y = Intercept + b1 × D + b2 × X1 + b3 × X2

INDEX

association, 10

bias, 37
   collider, 106–108
   confounding bias, 60
   unbiased causal estimates, 43

causal claims
   assessment, 109–114
causal discovery, 67
causal fallacy, 3
causal graphs
   see directed acyclic graphs
      (DAG), 6
causal inference, 5
causal language, 97–100
causal questions, 22
   forward, 23
   identifying, 97–100
   reverse, 23
causal study, 21
   assignment, 21, 22
   identifying, 97–100
   quasi-experimental study, 75
   random assignment, 42
   randomized controlled trials
      (RCT), 24
   unit (or subject), 22
   interventional study, 24
   observational study, 24
causality, 2
cause
   indirect, 62
cause and effect, 2
   outcome, 19
   treatment, 19
   causal statements, 17
   deterministic definition, 5
   probabilistic definition, 5
   words that imply causality, 18
chain, 62, 63
cinical trials
   see randomized controlled trials
      (RCT), 103
collider, 62, 64–66
   bias, 66, 106–108
conditioning on a variable, 62
conditioning on a variables, 62
confounders, 12, 60, 61
confounding bias, 12, 60
   identifying, 67
   observational study, 104
   unobserved, 71

control group, 21
controlling for a variable, 62
correlation
    vs. causation, 98–100
    correlation coefficient, 10
    correlational studies, 98–100

differences-in-differences
    parallel-trends assumption, 94
directed acyclic graphs (DAG), 6, 28–30, 33
    chain, 62
    collider, 64, 65
    colliders, 65, 66
    edge, 30
    edge (link), 28
    fork, 63
    how to construct, 67
    nodes, 28, 29
    outcome variable, 28
    paths, 31, 32, 34, 35
    treatment variable, 28
    cycle, acyclic, 30
    nodes, 29
    paths, 29

experimental study
    see randomized controlled trials (RCT), 103
exposure variable, 19
external validity, 85, 101, 102

First World War, 78
fork, 62–64

informed consent, 52
instrumental variables, 72
interventional data, 24
interventional studies, 40

longitudinal data
    see panel data, 86

natural experiments, 76–80
nodes, 29, 30
    dependent, 61
    independent vs. dependent, 61
    outcome node, 31
    treatment node, 31

observational data, 24
observational study, 59, 60, 104
    confounding bias, 104
    self-selection, 59
Oregon health insurance experiment, 79
outcome
    unit of measurement, 101
    variable, 19, 100

panel data, 86
path, 29
paths, 31
    blocking, 61
    causal vs. non-causal, 31, 32, 34, 35
    chain, 62, 63
    collider, 64–66
    non-causal, 48
    fork, 63, 64
placebos, 51
post hoc fallacy, 3

quasi-experimental study, 75–91
quasi-experiments, 86
questionable cause fallacy, 3

random assignment, 42
randomized controlled trials (RCT), 24, 40–43, 46, 48, 54, 103, 104
    attrition, 52, 103

randomized controlled trials
    (RCT) (*cont.*)
  blind study, 51
  ethical concerns, 50
  history of, 48–50
  informed consent, 51, 52
  non-compliance, 52
  shortcomings, 50
randomzied experiments
  see randomzied controlled trials
    (RCT), 109
regression analysis, 71
regression discontinuity, 81–86
  bandwidth, 85
  continuity assumption, 83
  sorting (bunching), 84
replication, 102
representative sample, 24, 101

sensitivity analysis, 72
Simpson's Paradox, 15
Spanish flu, 78
statistical significance, 104, 105
stratification, 70
swimmer's body illusion, 12

synthetic control, 86–90
  donor pool, 88
  limitations, 90, 91

treatment, 20
  average treatment effect (ATE),
    22, 46
  treatment effect, 22, 46
  treatment group, 21
  treatment variable, 19
  unbiased treatment effect, 53
  unit of measurement, 101
  variable, 100

variables
  binary, 21, 100
  categorical, 21
  confounders, 60
  continuous, 21, 100
  dependent, 61
  discrete, 21, 100
  independent, 61
  outcome, 19, 100
  running, 82
  treatment, 100